This report contains the collective views of an international group of experts and does not necessarily represent the decisions or the stated policy of the United Nations Environment Programme, the International Labour Organisation, the World Health Organization, or the Food and Agriculture Organization of the United Nations

Environmental Health Criteria 70

PRINCIPLES FOR THE SAFETY ASSESSMENT OF FOOD ADDITIVES AND CONTAMINANTS IN FOOD

Published under the joint sponsorship of the United Nations Environment Programme, the International Labour Organisation, and the World Health Organization in collaboration with the Food and Agriculture Organization of the United Nations

World Health Organization
Geneva, 1987

The International Programme on Chemical Safety (IPCS) is a joint venture of the United Nations Environment Programme, the International Labour Organisation, and the World Health Organization. The main objective of the IPCS is to carry out and disseminate evaluations of the effects of chemicals on human health and the quality of the environment. Supporting activities include the development of epidemiological, experimental laboratory, and risk-assessment methods that could produce internationally comparable results, and the development of manpower in the field of toxicology. Other activities carried out by IPCS include the development of know-how for coping with chemical accidents, coordination of laboratory testing and epidemiological studies, and promotion of research on the mechanisms of the biological action of chemicals.

ISBN 92 4 154270 5

©World Health Organization 1987

Publications of the World Health Organization enjoy copyright protection in accordance with the provisions of Protocol 2 of the Universal Copyright Convention. For rights of reproduction or translation of WHO publications, in part or *in toto,* application should be made to the Office of Publications, World Health Organization, Geneva, Switzerland. The World Health Organization welcomes such applications.

The designations employed and the presentation of the material in this publication do not imply the expression of any opinion whatsoever on the part of the Secretariat of the World Health Organization concerning the legal status of any country, territory, city or area or of its authorities, or concerning the delimitation of its frontiers or boundaries.

The mention of specific companies or of certain manufacturers' products does not imply that they are endorsed or recommended by the World Health Organization in preference to others of a similar nature that are not mentioned. Errors and omissions excepted, the names of proprietary products are distinguished by initial capital letters.

ISSN 0250-863X
PRINTED IN FINLAND
86/7121 — VAMMALA — 5500

CONTENTS

PRINCIPLES FOR THE SAFETY ASSESSMENT OF FOOD ADDITIVES
AND CONTAMINANTS IN FOOD

FOREWORD 11

PREFACE 12

1. INTRODUCTION 15

2. HISTORICAL BACKGROUND 18

   2.1 Introduction 18
   2.2 Periodic review 21
       2.2.1 Concept of periodic review 21
       2.2.2 Mechanism of periodic review 23

3. CRITERIA FOR TESTING AND EVALUATION 25

   3.1 Criteria for testing requirements 25
       3.1.1 Estimating exposure 26
       3.1.2 Predicting toxicity from chemical
             structure 27
       3.1.3 Other factors to consider when develop-
             ing criteria 28
   3.2 Priorities for testing and evaluation 29
   3.3 Quality of data 30

4. CHEMICAL COMPOSITION AND THE DEVELOPMENT OF
   SPECIFICATIONS 32

   4.1 Identity and purity 32
   4.2 Reactions and fate of food additives and
       contaminants in food 33
   4.3 Specifications 34

5. TEST PROCEDURES AND EVALUATION 39

   5.1 End-points in experimental toxicity studies . . . 39
       5.1.1 Effects with functional manifestations . . . 40
       5.1.2 Non-neoplastic lesions with morpho-
             logical manifestations 41
       5.1.3 Neoplasms 42
       5.1.4 Reproduction/developmental toxicity 46
       5.1.5 *In vitro* studies 48
   5.2 The use of metabolic and pharmacokinetic
       studies in safety assessment 50

			Page
	5.2.1	Identifying relevant animal species	51
	5.2.2	Determining the mechanisms of toxicity	53
	5.2.3	Metabolism into normal body constituents	54
	5.2.4	Influence of the gut microflora in safety assessment	56
		5.2.4.1 Effects of the gut microflora on the chemical	56
		5.2.4.2 Effects of the chemical on the gut microflora	59
5.3	Influence of age, nutritional status, and health status on the design and interpretation of studies		59
	5.3.1	Age	60
		5.3.1.1 History	60
		5.3.1.2 Usefulness of studies involving *in utero* exposure	64
		5.3.1.3 Complications of aging	65
	5.3.2	Nutritional status	66
	5.3.3	Health status	68
	5.3.4	Study design	68
5.4	Use of human studies in safety evaluation		70
	5.4.1	Epidemiological studies	72
	5.4.2	Food intolerance	73
5.5	Setting the ADI		75
	5.5.1	Determination of the no-observed-effect level	77
	5.5.2	Use of the safety factor	78
	5.5.3	Toxicological versus physiological responses	82
	5.5.4	Group ADIs	82
	5.5.5	Special situations	83
	5.5.6	Comparing the ADI with potential exposure	84
6.	PRINCIPLES RELATED TO SPECIFIC GROUPS OF SUBSTANCES		86
6.1	Substances consumed in small amounts		86
	6.1.1	Food contaminants	86
	6.1.2	Food flavouring agents	88
6.2	Substances consumed in large amounts		92
	6.2.1	Chemical composition, specifications, and impurities	93
	6.2.2	Nutritional studies	94
	6.2.3	Toxicity studies	95
	6.2.4	Foods from novel sources	97

		Page
REFERENCES	. . . .	100
ANNEX I.	GLOSSARY	110
ANNEX II.	STATISTICAL ASPECTS OF TOXICITY STUDIES . .	115
REFERENCES TO ANNEX II	. . . .	134
ANNEX III.	GUIDELINES FOR THE EVALUATION OF VARIOUS GROUPS OF FOOD ADDITIVES AND CONTAMINANTS	135
REFERENCES TO ANNEX III	. . . .	144
ANNEX IV.	EXAMPLES OF THE USE OF METABOLIC STUDIES IN THE SAFETY ASSESSMENT OF FOOD ADDITIVES	145
REFERENCES TO ANNEX IV	. . . .	152
ANNEX V.	APPROXIMATE RELATION OF PARTS PER MILLION IN DIET TO MG/KG PER DAY	156
ANNEX VI.	REPORTS AND OTHER DOCUMENTS RESULTING FROM MEETINGS OF THE JOINT FAO/WHO EXPERT COMMITTEE ON FOOD ADDITIVES	158
INDEX	. . . .	166

WHO TASK GROUP ON UPDATING THE PRINCIPLES FOR THE SAFETY
ASSESSMENT OF FOOD ADDITIVES AND CONTAMINANTS IN FOOD

1985 and/or 1986 JECFA Members

a,b,c,d,e	Dr H. Blumenthal, Division of Toxicology, Center for Food Safety and Applied Nutrition, US Food and Drug Administration, Washington DC, USA
c,d	Dr I. Chakravarty, Department of Biochemistry and Nutrition, All India Institute of Hygiene and Public Health, Calcutta, India
c	Dr W.H.B. Denner, Food Composition and Information Unit, Food Sciences Division, Ministry of Agriculture, Fisheries and Food, London, United Kingdom
c	Dr A.H. El-Sebae, Pesticides Division, Faculty of Agriculture, Alexandria University, Alexandria, Egypt
c	Professor P.E. Fournier, Hôpital Fernand-Widal, Paris, France
a,c,d,f	Dr S. Gunner, Food Directorate, Health Protection Branch, Health and Welfare Canada, Ottawa, Ontario, Canada
d	Mr J. Howlett, Food Science Division, Ministry of Agriculture, Fisheries and Food, London, United Kingdom
c,d,f	Professor K. Kojima, College of Environmental Health, Azabu University, Sagamihara-Shi, Kanagawa-Ken, Japan
c	Dr W. Kroenert, Food Chemistry Division, Max von Pettenkofer Institute, Federal Office of Public Health, Berlin (West)
a,b,c,d,f	Dr B. MacGibbon, Division of Toxicology, Environmental Pollution and Prevention, Department of Health and Social Security, London, United Kingdom
c,d	Dr R. Mathews, Food Chemicals Codex, National Academy of Sciences, Washington DC, USA
d	Mrs I. Meyland, National Food Institute, Ministry of the Environment, Soborg, Denmark
c,d,e	Dr J. Modderman, Food Additives Chemistry Evaluation Branch, Center for Food Safety and Applied Nutrition, US Food and Drug Administration, Washington DC, USA
d	Dr G. Nazario, Ministry of Health, National Health Council, Rio de Janeiro, Brazil
c,d	Professor K.A. Odusote, College of Medicine, University of Lagos, Lagos, Nigeria

1985 and/or 1986 JECFA Members (contd).

c,d	Professor F. Pellerin, Faculté de Pharmacie de l'Université Paris-Sud, Hôpital Corentin Celton, Issy-les-Moulineaux, France
c,f	Dr P. Pothisiri, Food Control Division, Food and Drug Administration, Ministry of Public Health, Bangkok, Thailand
b,c,d	Professor M.J. Rand, Department of Pharmacology, University of Melbourne, Parkville, Victoria, Australia
a,b,d,e,g	Dr P. Shubik, Green College, Oxford, United Kingdom
d	Dr A. Slorach, Food Research Department, The National Food Administration, Uppsala, Sweden
d	Dr V.A. Tutelyan, Institute of Nutrition, Academy of Medical Sciences of the USSR, Moscow, USSR

Secretariat

d	Dr Y.K. Al-Mutawa, Division of Public Health Laboratory, Ministry of Public Health, Safat, Kuwait
a	Dr E.A. Bababunmi, University of Ibadan, Ibadan, Nigeria
e	Dr A. Bär, Department of Vitamin and Nutrition Research, F. Hoffmann-LaRoche and Company, Ltd., Basel, Switzerland
a,b,d,f	Dr J. Cabral, Unit of Mechanisms of Carcinogenesis, International Agency for Research on Cancer, Lyons, France
e	Dr J. Caldwell, St. Mary's Hospital Medical School, London, United Kingdom
a,e	Dr D.M. Conning, British Nutrition Council, London, United Kingdom
f	Dr J.L. Emerson, External Technical Affairs, The Coca-Cola Company, Atlanta, Georgia, USA
d	Mr A. Feberwee, Committee on Food Additives, Nutrition and Quality Affairs, Ministry of Agriculture and Fisheries, The Hague, The Netherlands
d	Professor C.L. Galli, Toxicology Laboratory, Institute of Pharmacology and Pharmacognosy, University of Milan, Milan, Italy
e	Dr M.J. Goldblatt, Consumer Nutrition Affairs, General Foods Corporation, White Plains, New York, USA
f	Dr W. Grunow, Divison of Food Toxicology, Max von Pettenkofer Institute, Federal Office of Public Health, Berlin (West)
d	Mr R. Haigh, Commission of the European Communities, Brussels, Belgium

Secretariat (contd).

a,d,f	Dr Y. Hayashi, Division of Pathology, Biological Safety Research Center, National Institute of Hygienic Sciences, Tokyo, Japan
b,d,e,g	Dr J. Herrman, Division of Food and Color Additives, Center for Food Safety and Applied Nutrition, US Food and Drug Administration, Washington DC, USA
f	Dr D. Krewski, Environmental Health Directorate, Health Protection Branch, Health and Welfare Canada, Ottawa, Ontario, Canada
e	Mr P.N. Lee, Consultant in Statistics and Adviser in Epidemiology and Toxicology, Surrey, United Kingdom
b	Dr M. Mercier, International Programme on Chemical Safety, World Health Organization, Geneva, Switzerland
d	Dr R.W. Moch, Center for Food Safety and Applied Nutrition, US Food and Drug Administration, Washington DC, USA
a	Dr V.H. Morgenroth, Chemicals Division, Environment Directorate, Organization for Economic Cooperation and Development, Paris, France
a	Dr I. Nir, Department of Pharmacology and Experimental Therapeutics, The Hebrew University Hadassah Medical School, Jerusalem, Israel
d	Dr E. Poulsen, National Food Institute, Institute of Toxicology, Soborg, Denmark
b,d,f	Dr A.W. Randell, Food Policy and Nutrition Division, Food and Agricultural Organization of the United Nations, Rome, Italy
d	Dr N. Rao Maturu, Joint FAO/WHO Food Standards Programme, Food and Agricultural Organization of the United Nations, Rome, Italy
e,f	Dr A.G. Renwick, Clinical Pharmacology Group, University of Southampton, Southampton, United Kingdom
e,f	Dr F.J.C. Roe, Consultant in Toxicology and Adviser in Experimental Pathology and Cancer Research, London, United Kingdom
d,e	Dr S.I. Shibko, Division of Toxicology, Center for Food Safety and Applied Nutrition, US Food and Drug Administration, Washington DC, USA
a,b,e,f	Dr V. Silano, Department of Comparative Toxicology, High Institute of Health, Rome, Italy
d	Professor A. Somogyi, Department of Drugs, Animal Nutrition and Residue Research, Institute for Vetinary Medicine, Berlin (West)

Secretariat (contd).

b,d	Professor R. Truhaut, Faculté des Sciences Pharmaceutiques et Biologiques de Paris Luxembourg, Laboratoire de Toxicologie et Hygiene Industrielle, Université Rene Descartes, Paris, France
f	Dr G.J. Van Esch, National Institute for Public Health and Environmental Hygiene, Bilthoven, The Netherlands
a,b,d	Dr G. Vettorazzi, International Programme on Chemical Safety, World Health Organization, Geneva, Switzerland
e	Dr M.J. Wade, Division of Toxicology, Center for Food Safety and Applied Nutrition, Washington DC, USA
a,b,d,e,g	Dr R. Walker, Department of Biochemistry, University of Surrey, Guildford, Surrey, United Kingdom

---

[a] Present at strategy meeting, Oxford, United Kingdom, 19-23 September, 1983.
[b] Present at pre-consultation of contributors, Geneva, Switzerland, 29-31 May, 1985.
[c] Member of JECFA-85, Geneva, Switzerland, 3-12 June, 1985.
[d] Participant in JECFA-86, Rome, Italy, 2-11 June, 1986.
[e] Consultant who contributed written material.
[f] Submitter of written comments.
[g] Member of Editorial Committee.

NOTE TO READERS OF THE CRITERIA DOCUMENTS

Every effort has been made to present information in the criteria documents as accurately as possible without unduly delaying their publication. In the interest of all users of the environmental health criteria documents, readers are kindly requested to communicate any errors that may have occurred to the Manager of the International Programme on Chemical Safety, World Health Organization, Geneva, Switzerland, in order that they may be included in corrigenda, which will appear in subsequent volumes.

* * *

FOREWORD

The WHO activities concerned with the safety assessment of food chemicals were incorporated into the International Programme on Chemical Safety (IPCS) in 1980. Since this time, a keen interest has developed in all aspects pertaining to the toxicological evaluation of food additives and contaminants, including the methodological aspects. These activities are part of the responsibilities of the Programme insofar that its objectives include the formulation of "guiding principles for exposure limits, such as acceptable daily intakes for food additives and pesticide residues, and tolerances for toxic substances in food, air, water, soil, and the working environment".

The present publication on "Principles for the Safety Assessment of Food Additives and Contaminants in Food" has been developed in response to repeated recommendations by the Joint FAO/WHO Expert Committee on Food Additives (JECFA). Its inclusion in the methodology section of the Environmental Health Criteria series will make it readily available to both Member States and the food industry.

The IPCS gratefully acknowledges the financial support of the United Kingdom Department of Health and Social Security (DHSS), and the US Food and Drug Administration (FDA), which was indispensable for the completion of the project.

                        Dr M. Mercier
                        Manager
                        International Programme on
                            Chemical Safety

PREFACE

For the last thirty years, the internationally sponsored committee known as the Joint FAO/WHO Expert Committee on Food Additives (JECFA) has played a major role in providing a unique international mechanism for the identification and safety assessment of food chemicals, including food additives, food contaminants, and residues of veterinary drugs. With no regulatory aspirations, this Committee has probably contributed more to the elaboration of sound national food regulation in this area than any other international body aimed at harmonizing or normalizing often divergent national approaches to the problem of food safety, food technology, and food control. JECFA achieved this by providing recommendations based on scientific evidence and by establishing a rational model of safety assessment that is widely reputed and accepted.

Hundreds of highly skilled international specialists have given, and continue to give, freely of their time and talents to foster advances in toxicological methodologies and analytical procedures, to consolidate accessible presentations of data, and to keep abreast with scientific developments, which often requires readjustment of previous conclusions. Through reports, toxicological monographs, and profiles of chemical specifications by JECFA, national food regulatory authorities and the Codex Alimentarius Commission are provided with all the necessary elements for making the best decisions on the rational use of chemicals in food.

The present undertaking has several precedents in the history of JECFA. For example, in 1957, the second report of the Committee elaborated on procedures for the testing of intentional food additives to establish their safety for use and, in 1960, the fifth report contained a series of guidelines for the evaluation of the carcinogenic hazards of food additives. It should also be mentioned that, in 1966, the Committee commissioned a special scientific group to develop procedures for investigating intentional and unintentional food additives. Finally, in 1981, after realizing that a significant interval of time had elapsed since previous methodological updatings, the Committee called for a state-of-the-art review of methodology. A favourable answer was received from the newly established International Programme on Chemical Safety (IPCS), a cooperative programme sponsored by the International Labour Organisation (ILO), the United Nations Environment Programme (UNEP), and the World Health Organization (WHO). It should be noted that the implementation of the recommendation by the IPCS was significantly facilitated by the fact that the toxicological component of the JECFA came within the scope of the programme.

The contents of this publication are the result of sustained efforts of an IPCS Task Group during a number of meetings including: a strategy meeting in 1983, a consultation of contributors in 1985, and the JECFA Working Groups at their annual meetings in 1985 and 1986. The members of the Task Group contributed either written material or comments, or both. An Editorial Board was responsible for preparing the final draft for publication. The Task Group benefited widely from the large number of recommendations and observations on the methodology of testing and assessing chemicals in food, found in the previous reports of JECFA and related scientific groups.

Thus, this publication reflects faithfully the recommendations of the Committee regarding the safety assessment of food additives and contaminants by reaffirming the validity of recommendations that are still appropriate, while pointing out the problems associated with those that, in the light of modern advances in methodology, are no longer valid. New recommendations are also made, as might be expected, with the advancing state of the toxicological sciences. Particularly enlightening are the section dealing with the principles related to the safety assessment of substances consumed in large amounts, and Annexes II and IV dealing with the statistical aspects of toxicity studies and examples of the use of metabolic studies in the safety assessment of food additives, respectively. An Index has also been included at the end of the book.

It is the most earnest wish of all concerned with the production of this publication that it should make an important contribution to the field of food toxicology and that it will be found useful by the members of JECFA, national food regulatory authorities, and industry, who are involved in the development of safety data and in making the consumer aware of the problems of the safe use of food additives.

Dr G. Vettorazzi

Dr A. Randell

1. INTRODUCTION

This publication is concerned with reviewing the basis for decision-making by the Joint Expert Committee on Food Additives (JECFA) of the Food and Agriculture Organization of the United Nations (FAO) and the World Health Organization (WHO). Because the toxicological and chemical characteristics of food additives are the primary concern of the Committee, both aspects are dealt with in this publication, which has been prepared by WHO- and FAO-appointed consultants, assisted by the WHO and FAO secretariats. The twenty-eighth JECFA report (1) includes a summary of the areas that the Committee considered to be most in need of evaluation.

The major concerns of this monograph are with the testing of chemicals in food and with the evaluation of the test results. In keeping with the approach developed during the past 30 years by JECFA, the recommendations for test procedures and safety assessment are discussed in broad terms taking into account the latest scientific advances in the relevant fields. No effort has been made to provide instructions for test procedures. Differences in national approaches to toxicological evaluations exist, and some variations in the data submitted must be considered by JECFA in making their evaluations. However, certain basic data requirements are necessary in order to enable an expert committee to make sound judgements. The elements of this base line are included in various sections of this report.

In essence, the problems under consideration fall into three general categories: first, the determination of the chemical and toxicological test requirements for individual chemicals that are added to or occur in food; second, the assessment methods that are to be applied; and third, the updating of the test procedures and methods of assessment as the science progresses.

Methods for testing and assessment have changed considerably during the life of JECFA. However, the Committee has, by no means, been a static organization. Not only have various approaches been continuously updated at the individual meetings of the Committee, but intervening meetings of scientific groups have been held to consider the impact of new scientific developments on the procedures used (2, 3). Thus, every effort has been made in this publication to record the views of JECFA and to detail the changes that have come about with the course of time.

It is recognized that, with advances in science, it is possible to obtain more complete toxicological profiles for individual chemicals. For example, it is becoming increasingly easy to learn more about the disposition and metabolic fate of xenobiotics. Until relatively recently, the toxicologist in this field had to rely almost entirely on a set of routine tests, the results of which were then assessed and used to establish arbitrarily-determined safe levels.

JECFA has long recognized that a number of factors can be used to determine test requirements; these include the structure of the chemical, its natural occurrence in foodstuffs, its metabolic characteristics, and knowledge of its effects in man. However, systematic guidelines incorporating these factors have not been developed by JECFA. In recent years, some of the problems posed to JECFA have concerned the testing and evaluation of additives and food ingredients that are consumed in large amounts. Other key problems in the determination of the appropriate level of testing involve the largest group of food additives, the flavouring agents that are used, generally, at very low levels and that are often "nature identical" or derived from natural sources. Therefore, both "high consumption" substances and flavouring agents are discussed in detail in sections 6.1 and 6.2

There have been considerable changes in laboratory studies used in other areas of toxicology, which are yet to have a major impact on the evaluation of food additives. Of particular interest are the series of mutagenicity/clastogenicity tests, often, but not invariably, using sub-mammalian test organisms *in vitro*. The number, diversity, and uses of these tests have increased rapidly in the past decade. In general, such tests are effective for measuring an intended genetic end-point. How effectively these tests identify chemical carcinogens is much less clear. In the absence of a clear correlation with carcinogenicity, it is difficult to know how such tests should be interpreted and used in safety evaluations. Even though these *in vitro* tests may not be required for the evaluation of the safety of food additives, it is becoming more and more frequent for chemicals to be tested in this way for other reasons, e.g., to detect potential environmental or occupational hazards. Thus, JECFA may have to decide on the relevance of such information (section 5.1.5).

During the past decade, there has been a major increase in the number of chemicals tested routinely for chronic toxicity in standardized *in vivo* tests. Although these tests may not be designed to evaluate food additives, the results have to be carefully considered at JECFA meetings. Among the major problems that occur are those arising from results obtained when chemicals are administered to animals by routes other than the diet or drinking-water. For example, some years ago, JECFA faced the problem of assessing the significance of the induction of subcutaneous sarcomas at the site of injection of certain food chemicals into rodents. It was found that many substances, including inert plastics, could give rise to similar sarcomas on implantation. As a result, the Committee concluded that such findings could not be used in a definitive manner for assessing the safety of food additives (4, pp. 16-17). However, these findings cannot be totally ignored and may be an indicator of the need for further carcinogenicity studies using the oral route. Another problem with interpretation arises with many of the more recent "routine" *in vivo* studies that record the enhancement of a variety of common "spontaneous" tumours in rodents, including lymphomas, hepatomas, and pheochromocytomas (section 5.1.3).

The scope of long-term toxicity tests has been discussed extensively by JECFA. For example, several food chemicals have been tested in 2-generation studies rather than the commonly-used single-generation study. While the use of this more extensive test is advisable under certain conditions, it should not necessarily be a routine procedure (section 5.3).

Many of the chemicals of concern to those responsible for food safety evaluation are present in food at very low levels and may be present as environmental contaminants or may result from the migration of substances from food packaging or residues from the use of solvents, pesticides, or veterinary drugs. These situations often require very different approaches to test requirements than those used for intentional food additives (Annex III). One case, the use of anabolic agents in livestock, has posed various problems for JECFA that cannot be answered within the scope of current procedures (Annex III). JECFA will soon be developing methodology for the testing and evaluation of veterinary drug residues in support of a new committee on residues of veterinary drugs in foods that has been established by the Codex Alimentarius Commission.

In assessing the significance of data, a major issue to be resolved concerns the distinctions that should be made among different toxicological manifestations. The carcinogenic potential of chemicals has been emphasized in the past few decades to the exclusion of most other toxic end-points. There was a general consensus that chemicals found to be carcinogenic were not appropriate as food additives at any level whatsoever. More recently, however, it has become widely accepted that the term "carcinogen" has become harder and harder to define (section 5.1). It is apparent that cancer can be induced by a variety of chemicals acting by very different mechanisms and that the mechanism should be an important consideration when determining whether a safe level can be established. Other questions concern whether high-dose animal data are relevant to human exposure to low levels, and how teratogenicity data in the absence of maternal toxicity are to be interpreted (section 5.1.4).

2. HISTORICAL BACKGROUND

2.1 Introduction

The Joint FAO/WHO Expert Committee on Food Additives (JECFA) was established following recommendations made to the Directors-General of FAO and WHO by the Joint FAO/WHO Expert Committee on Nutrition at it fourth session (5), and the subsequent first Joint FAO/WHO Conference on Food Additives was held in September, 1955 (6). The terms of reference of the earlier meetings of JECFA related to the formulation of general principles governing the use of food additives and consideration of suitable uniform methods for evaluating the safety of food additives. For these purposes, food additives were defined by the Joint Conference as "non-nutritive substances added intentionally to food, generally in small quantities, to improve its appearance, flavour, texture, or storage properties."[a] Following recommendations of the third Joint FAO/WHO Conference on Food Additives (8), these terms of reference were broadened to include substances unintentionally introduced into human food and JECFA has subsequently considered and evaluated such materials, including growth promoters, components of packaging materials, solvents used in food processing, aerosol propellants, enzymes used in food processing, and metals in foods. Novel foods and ingredients that may be incorporated into foods at levels higher than those previously envisaged for food additives have also been referred to JECFA and pose special problems in safety evaluation, which will be discussed later in this report (section 6.2).

The first (9), second (10), and fifth reports (4) of JECFA established principles for the use of food additives and made recommendations on methods for establishing the safety-in-use of food additives and for the evaluation of the carcinogenic hazards of food additives. From the outset, JECFA recognized that:

"no single pattern of tests could cover adequately, but not wastefully, the testing of substances so diverse in structure and function as food additives" and that "the establishment of a uniform set of experimental procedures that would be standardized and obligatory is therefore undesirable" (10).

---

[a] From a practical standpoint, the "food additive" definition has been expanded since the time it was drafted, as a variety of compounds, including nutritive substances consumed in high amounts, have been brought under the umbrella of food additives. Indeed, the second Joint FAO/WHO Conference on Food Additives (7) recommended that the scope of the JECFA programme be expanded beyond the substances included in the original definition.

Accordingly, this Committee concluded "that it was only possible to formulate general recommendations with regard to testing procedures." The Committee also recognized that advances in the basic sciences might suggest new approaches to toxicological investigations and that these might be used immediately by the scientist but would take longer to become incorporated into any officially recommended testing procedures. Subsequent meetings of JECFA have consistently adopted this approach and have avoided the adoption of fixed protocols for the testing and evaluation of all classes of intentional and unintentional food additives. This has had the advantage of allowing the Committee to respond to new problems as they have arisen, with minimal inertia, and to encompass non-routine and ad hoc studies in the safety evaluation process. Within this framework, the Committee has found it possible to formulate guidelines for the evaluation of several groups of intentional and unintentional food additives that posed their own peculiar problems; several of these guidelines, which serve as specific examples to support general principles, are contained in Annex III.

The requirement to keep abreast of scientific developments in toxicology and related scientific disciplines implies the need for a periodic review of testing methodology. Following recommendations to this effect made by the eighth (11) and ninth (12) meetings of JECFA, a WHO Scientific Group on Procedures for Investigating Intentional and Unintentional Food Additives was convened in 1966

"to review, in the light of new scientific knowledge, the criteria used in establishing acceptable daily intakes. . . ." and "to suggest further studies on toxicological procedures used for the evaluation of intentional and unintentional food additives in order to establish their safety to the consumer" (2).

Subsequent meetings of JECFA have taken cognisance of the report of this Scientific Group and of the report of a more recent WHO Scientific Group on the Assessment of Carcinogenicity and Mutagenicity of Chemicals (3). Some aspects of the reports of other WHO Scientific Groups on the Principles for the Testing and Evaluation of Drugs for Carcinogenicity (13), Mutagenicity (14), and Teratogenicity (15) are also pertinent to the methodology of testing food additives. However, significant developments in the science of toxicology and related disciplines led the seventeenth meeting of JECFA to recommend that "the methods and procedures for the toxicity testing of food additives should be comprehensively reviewed and brought into line with advances in toxicology and cognate disciplines" (16). This recommendation was reiterated in the reports of the eighteenth (17) and nineteenth (18) meetings, the latter of which called for the convening of an appropriate meeting for the purpose of the review, and reaffirmed at the twentieth meeting (19).

Safety evaluation of food additives is a 2-stage process. The first stage involves the collection of relevant data including the results of studies on experimental animals and, where possible, observations in man. The second stage involves the assessment of data to determine the acceptability of the substance as a food additive. While the recommendations referred to in the preceding paragraph emphasize the impact of scientific advances on the first testing stage, the impact of such advances extends also to the assessment stage. This was made explicit in the twenty-first report of JECFA (20, p. 31), which stated that:

> "in view of the rapid progress of the science of toxicology and the increasing refinement of evaluation procedures, the Committee felt strongly that the traditional concepts of setting ADIs, the application of safety factors, and the relationship of these safety factors to the observed toxicological manifestations in animal experiments should be reconsidered".

This recommendation was endorsed by the twenty-fourth meeting of JECFA (21).

Many features of toxicity testing and evaluation of adventitious food additives and contaminants that fall within the terms of reference of JECFA, are common to pesticides that are within the scope of the Joint FAO/WHO Meeting on Pesticide Residues. In recognition of this, the twenty-fifth meeting of JECFA (22) recommended that a group of experts should be convened, as soon as possible, to study the application of advances in methodology to the toxicological evaluation of food additives and contaminants, and also of pesticide residues. The urgency of the need to implement this recommendation was stressed by the twenty-sixth (23) and twenty-seventh (24) meetings of JECFA.

In response to the Committee's repeated recommendations, a meeting of a group of experts to study the application of advances in methodology to the toxicological evaluation of food additives and contaminants was convened in September 1983. The objectives of the meeting were to formulate specific recommendations in order to bring up to date:

(a) the principles set out in earlier reports of JECFA concerning safety evaluation in relation to specific toxicological problems or specific chemical entities or groups;

(b) the test methods used in the toxicological evaluation of chemicals in food; and

(c) the assessment procedures adopted by JECFA in determining quantitative end-points.

The report of the Working Group (Updating Principles of Methodology for Testing and Assessing Chemicals in Food: Report of a Strategy Meeting) (unpublished WHO document ICS(Food)/83.3) and working papers on specific issues were considered by the twenty-eighth meeting of JECFA (1). Several questions were identified as remaining to be considered, including special problems associated with:

(a) bulking agents and novel foods;

(b) food contaminants;

(c) animal feed additives and veterinary drug residues;

(d) test methods and principles (including alternative methods of testing);

(e) testing for allergenicity;

(f) lesions observed in bioassays that are difficult to interpret (a number of examples are cited in the report); and

(g) assessment procedures; extrapolation and quantitative assessment.

The Committee recommended that a unified document on these issues should be prepared for consideration by JECFA at a future meeting. The present publication has been prepared in response to that recommendation.

In carrying out the review of methodology for the testing and evaluation of intentional and unintentional food additives, the working group has taken notice of recommendations, guidelines, and procedures adopted by national regulatory authorities and international/supra-national organizations including the Organization for Economic Cooperation and Development (OECD), the International Agency for Research on Cancer (IARC), and the European Economic Community (EEC) Scientific Committee for Food. It is recognized that, where possible, a unified approach should be adopted. However, the purposes for which these other bodies have formulated guidelines differ in detail from those of JECFA, and it is inappropriate to adopt these without modification to meet the needs of JECFA.

2.2 Periodic Review

2.2.1 Concept of periodic review

JECFA has indicated that, in discharging its duty to evaluate the safety-in-use of intentional and unintentional food additives, it may be necessary to carry out a periodic re-evaluation of substances previously assessed by the Committee.

The first JECFA meeting, in looking ahead, envisaged, in addition to the continuing evaluation of food additives, that there would be a re-evaluation process associated with the programme on food additive safety assessment (9). It stated:

"Permitted additives should be subjected to continuing observation for possible deleterious effects under changing conditions of use. They should be reappraised whenever indicated by advances in knowledge. Special recognition in such reappraisals should be given to improvements in toxicological methodology."

This principle was endorsed in the third (25), seventh (26), eighth (11), and ninth reports (12) of JECFA.

The "need for review of past recommendations" was highlighted in the thirteenth JECFA report as follows (27, p. 22):

"There is a widespread but fallacious belief that clearance of an additive for use in food constitutes an irrevocable decision. Such a view renders a grave disservice to the cause of consumer protection for it fails to recognize the need for regular review of all safety evaluations."

Periodic review of past decisions on safety is made necessary by one or more of the following developments (27):

(a) A new manufacturing process for the food additive.

(b) A new specification.

(c) New data on the biological properties of the compound.

(d) New data concerning the nature, or the biological properties, or both, of the impurities present in a food additive.

(e) Advances in scientific knowledge germane to the nature or mode of action of food additives.

(f) Changes in consumption patterns or level of use of a food additive.

(g) Improved standards of safety evaluation. This is made possible by new scientific knowledge and the quality and quantity of safety data considered necessary in the case of new additives. Since JECFA began the evaluation of food additives in 1956, the paucity of information available on many food additives has been such that assessments have often been difficult to make. Tests of too short duration, conducted with a very small number of animals at inappropriate dose levels, and without adequate clinical, haematological, chemical, or histopathological examinations have frequently been encountered among the data submitted for evaluation. Tests of this sort cannot be regarded as having permanent validity; with the passing of time, they need to be supplemented by studies carried out in full accordance with the recommendations set out in the report of the WHO Scientific Group on Procedures for Investigating Intentional and Unintentional Food Additives (2).

It should also be noted that the second Joint FAO/WHO Conference on Food Additives (7) recommended that it (JECFA) "should revise, if needed, the toxicological evaluation of all additives considered in previous meetings of the Expert Committee."

The seventeenth report of JECFA (16) reads, in part:
"The objective in assessing the toxicological data on food additives is to ensure their safety for the consumer on the basis of all the evidence available to the Committee at the time. Future results with present methods or with techniques yet to be developed will necessitate reassessments that may lead to changes in earlier decisions."

Other meetings reaffirmed the need to take advantage of recent developments in toxicological techniques for research and safety assessment.

These Committee recommendations and observations, the rationale set forth in the thirteenth report (27) and other reports for the need to review past decisions, and the ensuing years of progress in the science of toxicology and refinement of research, evaluation procedures, and changing consumption patterns all point towards the advisability of periodic review of this large class of substances.

2.2.2 Mechanism of periodic review

That a considerable amount of re-evaluation of substances is already carried out within the system is evident when the year-to-year agenda of JECFA is examined. Food additives are reassessed when new biological and chemical data are made available to FAO and WHO. In fact, new data are mandated on timetables established by JECFA when temporary ADIs are established (section 5.5.5). In addition, re-evaluations are made at the request of Member States and by the Codex Alimentarius Commission.

However, for many additives, the assessment has not been conducted using the more recently adopted procedures for investigating intentional and unintentional food additives. A review of past decisions also reveals that some additives have only had a cursory examination. The evaluation of these additives may have been based on limited data.

A periodic review programme on substances previously reviewed by JECFA should be constituted to reflect the changing state-of-the-art and to provide the best possible assurance to consumers of food additive safety. However, a mechanism has not yet been developed for the continuous systematic updating of safety information on food additives. Of course, even in the absence of a periodic review programme, if new data on a food additive raises suspicion of significant hazard, then immediate re-evaluation is conducted.

The use of an international forum to devise and implement a system for the periodic review of chemicals used in or on food and contaminants of food could also be of great economic and practical value to Member States. It would ensure a uniform approach to a complex toxicological problem, duplication of effort would be minimized, and emphasis on such a programme would give added reassurance to consumers throughout the world that the food supply continues to be safe. Perhaps such a programme could be developed in cooperation with the Codex Alimentarius Commission.

3. CRITERIA FOR TESTING AND EVALUATION

JECFA has always operated on the principle that testing requirements for all food additives should not be the same. Such factors as expected toxicity, exposure levels, natural occurrence in food (section 6.1), occurrence as normal body constituents (section 5.2), use in traditional foods, and knowledge of effects in man (section 5.4) should be taken into account. In relation to carcinogenic hazards, the Committee has stated that "the scope of the test required should depend on a number of factors, such as the nature of the substance, the extent to which it might be present in food, and the population consuming it" (4). More generally, the Committee has requested data on, *inter alia*, method(s) of manufacture, impurities, fate in food, levels of use of additives in food, and estimates of actual daily intake, and concluded that such information "was important and relevant both for the toxicological evaluation and for the preparation of specifications" (22). However, difficulties arise when an attempt is made to determine testing requirements because of problems in predicting toxicity, estimating levels of food additive use and natural occurrence, and obtaining human data. As discussed below, criteria for testing requirements can also be used to allocate priorities for the testing and evaluation of food chemicals.

3.1 Criteria for Testing Requirements

The establishment of principles for determining the appropriate amount of data that will be required to adequately evaluate the safety of additives at their estimated consumption levels is urgently needed to ensure consistency in decision-making by the Committee and to provide guidance to sponsors of food chemicals. Both exposure data and potential toxicity should be important considerations in the establishment of these principles. Consideration of only one of these elements to the exclusion of the other leaves serious deficiencies. If only exposure data are used, then no consideration is given to the wide range of toxicities observed among chemicals and no advantage is taken of the vast amount of bioassay data already in existence. If, on the other hand, only toxicity information, predicted or known, is used, then chemicals with known toxic properties or those related to chemicals of known high toxicity, particularly carcinogenicity, would automatically require the most data, with no consideration given to relatively low exposure levels.

3.1.1 Estimating exposure

For the purpose of this publication, exposure is defined as the total intake of a chemical substance by human beings. For the majority of substances evaluated by JECFA, the primary mode of exposure is through ingestion of the substance in the food supply.

Estimates of exposure used by Committees in previous years are of three general types: *per capita* estimates, estimates from dietary food intake surveys, and analytical values from market-basket/total-diet surveys. For a detailed discussion of the advantages and the use of these different types of estimates see reference no. 28.

The *per capita* approach is an estimated value that represents the exposure level if a food additive or contaminant were equally distributed across a population. For example, a *per capita* intake for a nation may be calculated by dividing the total yearly production volume, corrected for imports and exports, of a chemical used in food, within a nation, by the national population. Another form of *per capita* intake may be computed from a nation's *per capita* disappearance of a certain food commodity multiplied by the usual level of an additive or contaminant in the food commodity. These *per capita* intakes can be converted to daily intake per kilogram body weight.

In some countries, dietary surveys are performed on foodstuffs consumed by a representative group of individuals, within a national population, over a short period of time, e.g., 1 - 14 days. The intake of an additive or contaminant, per food type, can be calculated by multiplying the usual additive or contaminant level in each type of food by the dietary intake of the food. The intakes per food type can then be summed to derive a total additive or contaminant intake. An advantage of the dietary survey approach is that additive or contaminant intakes for selected subpopulations, such as different age groups or high-frequency consumers of certain foodstuffs, may be computed, depending on the specificity of the dietary survey.

When considering intakes computed by the dietary survey approach, the tenth meeting of JECFA (29, pp. 23-24) reaffirmed the validity of calculating the average daily intake of a food additive based on: (a) levels arising from good technological practice; (b) average consumption of foods containing the additive; and (c) average body weight. This Committee also noted at the time that, while data on average food consumption were available from many countries, high consumption data were available from only two countries. The Committee recognized a special need for determining how much of a food additive is likely to be consumed by groups that have a high level of consumption and strongly recommended that every effort should be made to obtain such information on food consumption.

The fourteenth meeting of JECFA (30) considered methodologies for computing additive intakes from dietary food surveys, and recognized the importance of experimental design so that collective data can be used for calculating reliable intakes on an individual basis. The Committee noted difficulties in common descriptors for food items, when information is gathered in surveys performed by different organizations, and in obtaining confidential information about food additive use from industry.

Market-basket surveys (also called total-diet surveys) involve analyses of representative diets for the usual level of additive or contaminant in the diet. The analyses may be performed on food mixtures or on individual foodstuffs. The selection of foods represents a normal diet for a certain population, such as the typical daily diet for a certain nation's average consumer. Market-basket surveys can be used for estimating the actual level of additive or contaminant in a selected total diet, which is of value for substances present in food at levels that are less than the amounts added. However, the difficulties of analyses usually restrict this approach to estimations of average intakes of contaminants in samples representative of the average dietary habits of a nation's general population rather than estimations of intakes for selected subpopulations. In this regard, data on certain contaminants in food are available from the Global Environmental Monitoring System (GEMS).

These procedures are useful for estimating exposure to food additives in individual countries. However, accurate estimation is much more difficult when attempted on a global scale. Clearly, consumption of a food additive will not be the same in two countries in which it is regulated with differing restrictions or with very different food consumption patterns. To use exposure estimates on such a scale as a criterion for testing requirements or for setting priorities for the testing of food additives is an extremely ambitious exercise that would require extensive resources.

3.1.2 Predicting toxicity from chemical structure

Chemical structure determines to a great extent the attitude of the toxicologist towards a compound. As a result, there have been many efforts to systematize the use of chemical structure as a predictor of toxicity. The use of such relationships has been suggested by JECFA with certain classes of flavouring agents (section 6.1.2), and chemical structure is an important consideration in the selection of compounds for carcinogenicity testing. Structure/activity relationships also form the basis for establishing group ADIs (section 5.5.4).

Structure/activity relationships appear to provide a reasonably good basis for predicting toxicity for some categories of compounds, primarily carcinogens, which are characterized by specific functional groups (e.g., nitrosamines, carbamates, epoxides, and aromatic amines) or by structural features and specific atomic arrangements (e.g., polycyclic aromatic hydrocarbons and aflatoxins). However, all these chemical groups have some members that do not seem to be carcinogenic or are only weakly so. Since structure/activity relationships are better established for carcinogens than for other toxic end-points, dependence on such predictions emphasizes suspect carcinogens at the expense of other forms of potential toxicity. However, as more chemicals are tested for toxicity and other end-points are identified in the future, the data base will become larger, which should permit more valid comparisons between structure and toxicity among more classes of compounds.

In terms of carcinogenic substances, another system that has sometimes been used for predictive purposes is a battery of tests for genotoxicity (possible applications of such tests are discussed in section 5.1.5).

3.1.3 Other factors to consider when developing criteria

The value of structure/activity relationships and exposure data in determining the extent of testing required may be considerably enhanced by collateral information on metabolism and pharmacokinetics. It has been previously accepted that:
"if a series of chemical analogues can be shown to give rise to the same metabolic product. . . it may be sufficient to carry out toxicological studies on a suitable representative of the series" and "where adequate biochemical and toxicological data on closely related chemicals are available, the objective (of toxicity tests) becomes the detection of any deviation from the established pattern. This can usually be determined by intensive studies of a few months duration when these are adequately designed and evaluated" (2).

More recently (31), JECFA has concluded that:
"if the chemical structure of a compound under consideration did not closely resemble that of any known toxic or carcinogenic compound, and, if the toxicological data on it, its metabolites, and its homologues did not give any cause for concern, these less extensive toxicological data might be used for the evaluation of the compound. . . . In the evaluation of a series of structurally-related compounds, complete toxicological data should be available for at least one member of the series. Other compounds in the series should be evaluated on the basis of these data, plus data on their natural occurrence and metabolism, and on the toxicology of their homologous compounds."

These principles can form the basis for determining the limited amount of testing that may be required for compounds that are closely related structurally. If the toxicological data base is adequate for the homologous compounds and suggests a low intrinsic toxicity, metabolic and pharmacokinetic data alone may be sufficient to make an evaluation of a related compound.

The results of studies on absorption, distribution, and metabolism may either increase or decrease the health concern from the use of the additive. For example, a relatively non-toxic additive may be transformed by liver enzymes into a substance with a much greater toxic potential, or *vice versa*. Correlations between structure and activity will often automatically include these considerations, because substances of a particular class will often be absorbed, distributed, and metabolized in similar ways. However, this will not always be the case, and these parameters should be specifically considered when making such correlations.

Other factors influence the extent and type of testing required for safety assessment. For example, the need for extensive testing may be mitigated when the substance occurs naturally in food and has a history of human use or when it is metabolized into normal body constituents (section 5.2.3). More extensive testing in animals may be necessary when the additive will be used in special populations at risk, such as pregnant women and very young infants (section 5.3). Human testing may be needed if problems of intolerance arise (section 5.4.2). The types of end-points, as discussed in section 5.1, must be considered in any criteria system that is established.

The development of criteria for determining the extent of required testing is worthy of extensive future study. Its value would be considerably enhanced by including it in the context of a priority-setting scheme, as discussed below, because additives should not be considered in isolation from one another.

3.2 Priorities for Testing and Evaluation

The primary basis for establishing the list of substances to be considered by JECFA is the recommendations of the Codex Committee on Food Additives (CCFA) and Member Governments. However, Committees have recognized the need for the establishment of a "priority list as a means of selecting the most relevant compounds for future evaluation. In order to establish priorities for the toxicological testing and evaluation of intentional and unintentional food additives", JECFA recommended at the twenty-second (32) and twenty-third (31) meetings that:

"FAO and WHO convene an inter-disciplinary group of experts to establish an inventory of compounds that have not yet been fully evaluated and to classify them in terms of their potential hazard to health on the basis of toxicological knowledge and extent of use."

The Committee has recognized that the most obvious need for allocating priorities is for the testing and safety evaluation of food flavouring agents (19). Committees continue to stress the need to establish priorities for testing and evaluating food additives (24,33).

One basis for establishing priorities for testing food additives is by using an index based on exposure levels and predicted toxicity. For examples of approaches using these parameters, see references 34-38.

As discussed in section 3.1.1, valid exposure data are extremely difficult to develop. However, comparative levels of consumption may be sufficient for the purpose of setting priorities for the testing of food additives. Therefore, even though accurate consumption estimates of wide geographical relevance will probably never be achieved, the lesser requirement of comparative estimates may be achievable to the extent necessary for JECFA's use, if the Committee decides to develop the information.

Because of the semiquantitative nature of much of the biological data available for predicting toxicity, rigorous analytical or statistical interpretation is not always possible. Therefore, expert interpretation and evaluation of the data, a time-consuming and expensive procedure, must be integrated into any automated decision-making mechanism that is developed. To ensure maximum usefulness, the priority-setting system should take account of all available toxicity and other biological information, including metabolic and human data. A properly-devised system will be capable of considering new data and can be modified using modern data processing methods and equipment.

3.3 Quality of Data

In recent years, various national regulatory agencies and international bodies have instituted codes of Good Laboratory Practice (GLP), the aim of which is to help underwrite the validity of studies by ensuring that they can be verified and reproduced. GLP codes require the maintenance of certain records regarding the performance of studies, including data from chemical and toxicological tests, which help ensure full documentation of the conduct and results of studies. However, GLP codes are not a substitute for scientific quality; an inappropriate study may be conducted according to GLP standards. On the other hand, a study that does not meet GLP criteria may still be scientifically sound.

JECFA has always judged studies on their merits, the main criteria being that the study was: (a) carried out with scientific rigor, and (b) reported in sufficient detail to enable comprehensive evaluation of the validity of the results.

Usually, studies that are published in the scientific literature are subjected to peer review prior to publication, and after publication, the results are open to refutation or confirmation. Unpublished reports, on the other hand, are not necessarily subjected to this scrutiny. For this reason, JECFA has repeatedly recommended that data brought before it be published (10, p. 6; 12, p. 7; 39, p. 7; 40). However, in point of fact, JECFA does review many high-quality studies that remain unpublished for proprietary and other reasons. Also, the Committee often requests unpublished raw data when published reports do not include sufficiently detailed data for an adequate safety review. Studies performed in compliance with GLP codes provide added assurance that the quality of unpublished data is acceptable. For these reasons, it is appropriate that JECFA experts continue to consider all the data brought before them, published, and unpublished, and make decisions about the validity of these data on an *ad hoc* basis. This means that the studies reviewed by the Committee should be fully documented.

4. CHEMICAL COMPOSITION AND THE DEVELOPMENT OF SPECIFICATIONS

The proper safety evaluation and use of food additives requires that they be chemically characterized. Therefore, the Committees review data relating to the identity of additives, impurities that may be present, and possible reaction products that may arise during storage or processing. "JECFA specifications" are then elaborated, taking these and other factors into account.

4.1 Identity and Purity

To establish the chemical identities of food additives, it is necessary to know the nature of the raw materials, methods of manufacture, and impurities (22). This information is used to assess the completeness of analytical data on the composition of additives and to assess the similarity of materials used in biological testing with those commercially produced. From information on raw materials and methods of manufacture, potential impurities in commercially manufactured chemical materials due to carry-over of contaminants in raw materials and by-products of the manufacturing process can be predicted.

To evaluate biological testing data from multiple studies, JECFA must have information on the chemical composition of the tested materials, which necessitates manufacturing information. Analytical data on the chemical composition of materials used in biological testing should be more detailed than a standard presentation of chemical specifications. Furthermore, materials used in biological testing should be representative of substances manufactured by actual commercial processes so that the materials administered to experimental animals will represent those ingested by consumers.

A food additive may be a single chemical substance, a manufactured complex chemical mixture, or a natural product. The need for complete information on chemical composition, including description, raw materials, methods of manufacture, and analyses for impurities, is equally valid for each type of additive. However, implementation of the requirement for chemical composition data may vary depending on the type of substance. For additives that are single chemical substances, it is virtually impossible to remove all impurities in their commercial production; therefore, analyses are generally performed on the major components and predicted impurities, with the highest significance placed on potentially toxic impurities. For commercially manufactured complex mixtures, such as mono- and diglycerides, information is needed on the range of substances commercially produced, with emphasis on descriptions of manufacturing processes, supported by analytical data on the components of the different commercial products. Natural products present particularly difficult problems because of their biological variability and because the chemical constituents are too numerous for regular analytical determinations; thus, the analyst is starting with an "unknown". For additives derived from natural products, it is vital that the sources and methods of manufacture are precisely defined. Chemical composition data should include analyses for general chemical characteristics, such as proximate analyses for protein, fat, moisture, carbohydrate, and mineral content, and analyses for specific toxic impurities carried over from raw materials or chemicals used in the manufacture of the substance. Further information necessary for the evaluation of "novel foods", which are usually substances derived from natural products, is provided in sections 6.2.1 and 6.2.4.

4.2 Reactions and Fate of Food Additives and Contaminants in Food

Biological testing of food chemicals must relate to their presence in food as consumed. This is an important consideration when added substances undergo chemical change in food. Therefore, data are necessary on the reactions and fate of additives or contaminants in food and their effects on nutrients (22).
Certain food additives perform their functional effect by reaction with undesirable food constituents (e.g., antioxidants react with oxygen in food and EDTA reacts with trace metals) or by reactions that modify food constituents (e.g., potassium bromate reacts with dough constituents). Food additives may also degrade under certain conditions of food processing, though such degradation is detrimental to their functional effect. For example, the sweetener aspartame is transformed to a diketopiperazine derivative at rates varying with the acidity and the temperature of the food. In previous evaluations of "reactive" additives, Committees have evaluated analyses for additive reaction products in food, as consumed, and biological testing data on either specific reaction products or samples of food containing the reaction products as consumers would ingest them.
For all intentional food additives proposed for evaluation, Committees request submission of four types of data related to reactivity:

(a) the general chemical reactivity of the additive;

(b) its stability during storage and reactions in model systems;

(c) reactions of the additive in actual food systems; and

(d) the additive's fate in living systems. These data are important for relating biological test data to the actual use of the additive in food.

Processing aids are substances that come into contact with food during processing and may unintentionally become part of food because of their incomplete removal. Committees have evaluated a number of processing aids, such as extraction solvents and enzyme preparations, for their safety in use. When evaluating a processing aid, information should be provided on its use and either analytical data on the amount of the processing aid carried over into food or a computed estimate of the amount to be expected in food. In some cases, a component of the processing aid may have the greatest potential for biological effects, such as ethylenimine leaching from polyethylenimine, an immobilizing agent used in the preparation of immobilized enzyme preparations.

Contaminants in food evaluated by previous Committees include environmental contaminants and substances migrating from food packaging. Of environmental contaminants, metals have been considered the most often. Committees request information on the chemical forms of metals in the food supply (e.g., ionic form and/or covalently bonded chemical form) and their concentration distribution in the food supply, as determined by analyses of food or experimental models for carry-over from environmental sources. For contaminants derived from food packaging, data are required on the identification of chemicals migrating from the packaging material and concentrations in food (analysed or estimated from migration modelling studies).

4.3 Specifications

Specifications are a necessary product of Committee evaluations, the purposes of which are 3-fold:

(a) to identify the substance that has been biologically tested;

(b) to ensure that the substance is of the quality required for safe use in food; and

(c) to reflect and encourage good manufacturing practice.

The first Joint FAO/WHO Conference on Food Additives (6) established a programme for the collection and dissemination of information on the chemical, physical, pharmacological, toxicological, and other properties of individual food additives. The first two meetings of the Joint Expert Committee, in preparing reports on "General Principles Governing the Use of Food Additives" (9) and "Procedures for the Testing of Intentional Food Additives to Establish Their Safety for Use" (10), recommended that specifications should be prepared, citing the need for:

(a) limiting impurities in food;

(b) identifying materials used in toxicity testing; and

(c) ensuring that the additive tested is the additive used in practice.

The third meeting of JECFA was devoted in its entirety to developing principles governing the elaboration of specifications and developing provisional specifications for the first group of additives evaluated by the Committee (25).

JECFA specifications are minimum requirements for the composition and quality of food-grade additives, allowing for acceptable variation in their production (18). These specifications are meant to be used internationally and to the extent that data are available, specifications are elaborated to cover suitable products manufactured in various parts of the world. The third meeting considered the value of specifications with regard to protection for the consumer, advice to regulatory organizations, standards for the food industry, and establishment of safety for use (relative to identification of materials used in biological testing in comparison with materials produced for commercial use). The format for specifications established by this meeting continues to be used in current JECFA specifications, that is, the additive is identified by synonym, definition (chemical name, formula, relative molecular mass, etc.), and description, its functional uses are listed, tests of identity and impurities are provided, and an assay for the major component(s) is provided. The third meeting of JECFA, recognizing that practical specifications could not specify every impurity, limited the scope of impurity tests to constituents of commercially produced substances that: (a) were related to the safe use of the additive; (b) might affect the usefulness of the additive; or (c) would serve as an indicator of good manufacturing practice. Finally, the meeting concluded that, for specifications to be acceptable on a global basis, they must be subject to continuing review and evaluation to take into account the presentation of new information, particularly with respect to different manufacturing processes and improved analytical methods.

In detailing the purposes of JECFA specifications, Committees have, through the years, refined the scope of their specifications and provided advice on how they should be used.

Specifications developed by each Committee should be read in conjunction with the report of that Committee. JECFA specifications apply to the material(s) that was toxicologically reviewed and take into account the uses of the additive (17, 41). Periodic review of specifications is required, because of changes in patterns of additive use, in raw materials, and in methods of manufacture. Comments on JECFA specifications by national and international organizations are valuable sources of information for periodic review (18, 27, 29).

JECFA specifications in their entirety describe substances of food-grade quality, and as such, they are directly related to toxicological evaluations and to good manufacturing practice. However, though specifications may include criteria that are important for commercial users of additives, they do not include requirements that are of interest only to commercial users (42).

Differences may exist between specifications prepared by national and international organizations; however, the Committee is not aware of any information indicating that these differences incur health risks for consumers (23). JECFA specifications are meant to be minimum requirements for the safe use of additives, and not every component of commercially manufactured chemical substances is subject to an impurity test (11). Test requirements in JECFA specifications are sufficient to ensure the safe use of commercially manufactured food additives. Substances of higher chemical purity (e.g., analytical grade reagents) are not excluded from use in food, even though such substances may deviate somewhat from the identification tests in the specifications, provided that they meet the stated requirements for specified purity tests and are otherwise suitable for use as food additives (18).

Since 1956, the meetings of JECFA have designated specifications as either full or tentative. Specifications were given the tentative designation from the third to twenty-second meetings because either the chemistry data needed to prepare specifications were not adequate or a temporary ADI was assigned to the additive. At, and since, the twenty-third meeting of JECFA, the tentative designation has been assigned only when the data necessary for preparing specifications were insufficient.

JECFA policy has been to prepare specifications for substances added to food, whenever constituents of the substance had the potential to be present in food. Initially, specifications were prepared only for intentional food additives that were added directly, to accomplish a functional effect in food. The fourteenth JECFA (30) evaluated extraction solvents, the first group of "processing aids" that had been reviewed by JECFA. This Committee concluded that, although extraction solvents are substantially removed from food, evaluation of the conditions of safe use of these solvents depends on the identity and purity of the solvents. Therefore, JECFA specifications were prepared. Since then, specifications have been reported for other processing aids such as antifoaming/defoaming agents, clarifying agents, decolourizing agents, enzyme preparations, filtering aids, packing gases, propellants, lubricants/release agents, odour/taste-removing agents, and yeast "food" (yeast nutrients). The twenty-seventh JECFA (24) decided that chemical reagents used in the preparation of food additives or processing aids (such as glutaraldehyde in the preparation of immobilized enzyme preparations or acetic anhydride in the manufacture of modified starches) do not usually need specifications. Carry-over of these reagents or their contaminants into food may be controlled in the specifications for purity of the additive or processing aid.

Food additives may be marketed as formulated preparations, such as a mixture of a main ingredient with a solvent vehicle and emulsifier. Specifications refer only to each ingredient in the formulated preparation as individual commercially-manufactured food additive substances. Mixtures should not be formulated in such a way that the absorption or metabolism of any ingredient is altered so that the biological data are invalidated (12, 42). Added substances such as anticaking agents, antioxidants, and stabilizers may also influence the results of tests given in specifications. Therefore, in its nineteenth report, JECFA recommended that manufacturers of food additives should indicate the presence of such added substances (18).

In considering whether specifications apply to food additive quality as manufactured or as received, JECFA has decided to prepare specifications to cover the normal shelf-life of the product. Limits are set for decomposition products that may form during normal storage. Manufacturers and users of food additives should ensure good packaging and storage conditions and use good handling practices to minimize deleterious changes in quality and purity (18). Information on changes in the composition of food additives during storage should be submitted for evaluation by the Committee.

In addition to periodic reviews to examine the consistency of specifications within classes of similar additives, JECFA periodically reviews specification test methods to update the analytical methodology of specifications. Two summaries of specification test methods have been published (43, 44), which provide guidelines for the application and interpretation of specification requirements and test methods. JECFA has made considerable progress in adopting modern analytical methodology for specification tests, whenever equipment and other supplies needed to perform the tests are accessible on a world-wide basis. However, because JECFA specifications are elaborated for world-wide use, certain analytical methods involving recently-developed techniques or equipment cannot be included until such techniques are available on an international scale. Alternative methods of analysis can be used to test products for conformity with specifications, provided that the methods and procedures used produce results of equivalent accuracy and specificity.

In order to foster international agreement on specifications for food-grade substances, JECFA seeks comments from member countries and international organizations. The Codex Alimentarius Commission systematically provides comments on JECFA specifications through the Codex Committee on Food Additives (CCFA) and endorses certain JECFA specifications as "Codex Advisory Specifications". The systematic review of JECFA specifications by the CCFA has provided JECFA with valuable data on novel manufacturing processes, previously unknown impurities, updated methodology, and advice on the format of JECFA specifications.

Although JECFA specifications and those of the Codex Alimentarius Commission are elaborated for many of the same purposes, the interpretation of these purposes may result in differences in specific requirements or test methods for the same food-grade substance. In replying to suggested changes in JECFA specifications from the CCFA or other interested parties, it may be decided to amend existing specifications, providing that the requested changes do not significantly lessen the assurance of food-grade quality embodied within the JECFA specifications and that the requested change conforms with the principles for elaboration of specifications established at previous meetings. A requested change in an existing full JECFA specification must be supported by scientific data.

5. TEST PROCEDURES AND EVALUATION

5.1 End-Points in Experimental Toxicity Studies

There are virtually no findings in experimental toxicology that can be simply extrapolated to man without careful thought. During the past two decades, there has been an increasing emphasis on carcinogenesis as a manifestation of chemical toxicity. Most of the other manifestations of chemical intoxication, for example, immunosuppression, have, by comparison, received relatively little attention. This has resulted in an unbalanced approach by the toxicologist in which the emphasis on end-points bears a less and less obvious relationship with disease patterns in man. For example, a survey of the recommendations for the evaluation of food additives has revealed that little or no attention is paid to the detection of cardiovascular lesions, even though these lesions are the most common cause of fatalities in the human population in developed countries. In addition, certain lesions commonly found in rodents that are the primary targets of the toxicologist do not have any counterpart in man. It seems reasonable that an effort should be made to relate toxicological findings more carefully to the human situation, recognizing that this will be a long-term project. In the meantime, when conducting experimental animal tests, special attention should be paid to alterations that indicate a potential for the test compound to adversely affect the cardiovascular, immunological, reproductive, or central nervous systems. If such alterations are detected, they should be investigated further using special studies aimed at clarifying their significance.

The end-points discussed in this section have been grouped for convenience into effects with:

(a) functional manifestations only;

(b) non-neoplastic morphological characteristics;

(c) neoplastic manifestations; and

(d) reproduction/developmental manifestations.

In view of the large number of effects encountered, it is possible to summarize only some of the specific observations. However, situations that have become controversial are dealt with in more detail.

Finally, this section concludes with a short discussion on the role of short-term *in vitro* tests in the safety assessment of food additives.

5.1.1 Effects with functional manifestations

Generalized weight loss, although having causes that are not solely physiological (section 5.5.3), does not necessarily involve any particular pathological lesions (section 5.5.3). Reduced weight gain has played a major role as an end-point in toxicological determinations in various ways. In a sense, it has often been used for determining various empirical indices in the absence of other manifestations. The procedures followed by JECFA for determining an ADI demand that a no-observed-effect level should be established. For this level to be established, it is necessary to establish an effect level and, when all else has failed, a generalized decrement in weight gain has been used for this purpose, provided reduced food intake is not the obvious cause. The other major use of a decrease in weight gain has been in establishing a maximum tolerated dose (MTD) (for a definition of the MTD, see Annex I).

Among the commonest effects observed in studies on food additives is a laxative effect; the physiological reasons for this are usually quite apparent and can be taken into account when considering the appropriate levels of use of additives causing this effect. In most instances, additives have been permitted that cause laxation at high levels in man when they have been otherwise non-toxic and can be used effectively at levels at which laxation does not occur.

Although a great deal has been said about the need to evaluate food additives and contaminants for the induction of possible behavioural changes, JECFA has hitherto devoted little time to evaluating such changes. Since it has been suggested that certain food constituents can produce behavioural changes in man, JECFA will, in the future, undoubtedly have to consider such effects. Unfortunately, good animal models have not been developed, and objective human data are difficult to obtain. It is not possible to recommend a simple series of tests at this time, primarily because there is no clear consensus on the kinds of studies that should be performed nor on the interpretation of the results.

Intolerance to food additives should always be considered a possibility, even though tests for reactions to food additives are not part of the normal data package that JECFA considers when assessing new food additives. Even when evidence of widespread intolerance to a food additive appears, it may still be very difficult to determine a cause-and-effect relationship. Problems include the often-anecdotal nature of much of the evidence, psychological factors, and problems with developing an adequate central data collection system. These points are discussed in more detail in section 5.4.

5.1.2 Non-neoplastic lesions with morphological manifestations

A number of lesions are relatively frequently observed in *in vivo* studies, particularly in rodents, that often give rise to controversy.

Non-specific liver enlargement or hypertrophy has been discussed by a WHO scientific group (2, pp. 18-19). In the past, this occurrence has been considered to be a manifestation of toxicity. More recently, it has been realized that this can often be a physiological response involving the induction of microsomal enzymes in the detoxification process that is reversible on removal of the test compound.

The formation of calculi in the urinary bladder is a frequent finding in rodent studies. Often, the formation of calculi may be followed by the formation of bladder tumours. It is not uncommon to find calculus formation in one rodent species and not in another. Under these circumstances, the nature of the calculi can sometimes be associated with specific metabolic changes that have led to their formation. This, in turn, may allow for a scientifically-based extrapolation to man, providing that human clinical studies are possible.

Caecal enlargement, a common finding in rodent studies, is a normal finding in rodents maintained on standard laboratory diets under germ-free conditions. It is also a common response of rodents, especially rats, to diets that include non-nutrient substances (e.g., certain permitted food colours and saccharin) or certain nutrients in excessive concentrations (e.g., modified starches, plant gums, lactose, and various polyols).

Most of the enlargement is attributable to increased luminal contents; in addition, the weight of the caecal wall after washing out the luminal contents is usually marginally more than normal. In haematoxylin- and eosin-stained sections, the caecal wall shows no remarkable features, and there is no evidence that caecal enlargement predisposes to any form of neoplasm of the caecum. Caecal enlargement may be due to osmotic effects, but its mechanism is not well understood. In some cases, caecal enlargement is an incidental finding, with the primary effect being nephrocalcinosis (23, pp. 11-12). Various forms of mineral deposition occur in the kidneys of laboratory rodents, more commonly in rats than in mice or hamsters. Unless appropriate diagnostic staining or chemical analysis is carried out, it is not strictly justifiable to refer to these changes as renal "calcinosis", though most of the mineral deposits do, in fact, contain calcium in one form or another. Mineral deposition can take place in almost any part of the nephron and deposition may predominate in one or more areas of the kidney. The main forms of renal mineralization are basement membrane mineralization, corticomedullary nephrocalcinosis, pelvic nephrocalcinosis, and nephrocalcinosis associated with acute tubular nephrosis. All these forms of nephrocalcinosis may co-exist, and one and the same agent may cause more than one form of nephrocalcinosis.

Magnesium deficiency in standard laboratory diets undoubtedly contributes to the high incidence of corticomedullary nephrocalcinosis in rats. Excessive dietary phosphate and possibly excessive dietary calcium may predispose to pelvic nephrocalcinosis. Such observations lead to the conclusion that more attention needs to be paid in the future to the formulation of diets for rodents with respect to physiologically-relevant levels of calcium, magnesium, and phosphorus.

Testicular atrophy, which is sometimes observed in rodents, may occur as a result of reduced caloric intake, frequently as a result of the addition of an unpalatable chemical to the animal's food or drinking-water. This should be distinguished from testicular atrophy resulting from the direct action of the chemical on the testicular cells. This distinction can be achieved by undertaking paired feeding studies. It is important that the function of the testes be investigated in reproduction studies when atrophy is detected.

Manifestations of vitamin deficiencies, notably of the fat-soluble vitamins, are sometimes observed in studies on agents that may be fat solvents and are only partly absorbed in the gastrointestinal tract; an example of such a substance is mineral oil. Another effect that is sometimes observed is discolouration of mesenteric lymph nodes after feeding a coloured substance. This is a normal physiological response and should not be considered a toxic end-point, as long as it is not associated with proliferative reactions.

Hormonally-associated effects occur with certain additives and require special endocrinological evaluations. Recently, JECFA has been faced with the task of evaluating the use of certain anabolic agents used in the raising of meat-producing animals (22, 23). These agents result in the presence of low-level residues of hormonally-active compounds in meat. The evaluation of the potential toxic effects of such compounds requires knowledge of the levels of naturally-occurring compounds with similar effects (Annex III).

5.1.3 Neoplasms

The most difficult decisions facing the toxicologist arise from the varied end-points found in long-term *in vivo* carcinogenesis studies. These problems have existed for a long time, but they have been greatly exacerbated by the large number of effects observed in the many rote tests now performed on chemicals including many that may be present in the food supply. These chemicals include certain direct food additives, some processing and carrier solvents, components of packaging materials, and a variety of contaminants.

Oncologists now generally recognize that different mechanisms of carcinogenesis exist with different chemicals acting on different tissues in the body (45, 46). Many believe that it may be possible to determine tolerance levels for some carcinogens, though this is still not possible with any degree of certainty with the majority of them. The view that a tolerance may be set for carcinogens giving rise to tumours through either a hormonal mechanism or by the formation of bladder calculi has been expressed by a WHO scientific working group (3, p. 11).

The perception that a chemical for which there is evidence of "carcinogenicity" in any system should not be permitted for use as a food additive at any dose whatsoever has become widespread. Although this philosophy has never been promulgated or officially adopted by JECFA, it has, in practice, influenced the Committee's approach to decision making. Probably more experimental work has been undertaken in cancer research over the past two decades, since this view was first established, than in all the preceding years, and clearly there is a great need to clarify the issue in terms of practical interpretation.

The assessment of the evidence for the carcinogenicity of chemicals is a major issue to be resolved by JECFA. Not only have many chemicals been tested by a variety of routes of administration that may not be relevant to food additive use (such as the repeated injection of food colours and other additives in rats and mice with consequent development of subcutaneous sarcomas at the site of injection (27)), but, in addition, new end-points are continually revealed and interpretation becomes more confusing as studies become more and more detailed. Positive results may be obtained due, for example, to a carcinogenic impurity. The extrapolation of such data has become very complex. One possibility, to make the term "carcinogen" more generalized, clearly would not solve the problem of how best to interpret these data.

Much of this issue centres around the meaning of the various types of enhancement of the tumours that occur in the rodents used for *in vivo* bioassays, since the rodents in common use, the mouse and the rat, develop extremely high incidences of a variety of tumours in the untreated state. Many reports indicate that one or more of these tumours has an increased incidence or has appeared earlier (or both) in treated, compared with untreated animals. One problem in interpreting the significance of these tumours is the difficulty in deciding whether these naturally-occurring tumours are spontaneously induced or whether the agent is able to induce them. The problem is further confounded by the fact that the incidence of tumours in the untreated control animals varies considerably with time. As a result, there is now a debate as to the importance of historical as well as concurrent control animals. It appears without doubt that both such controls are of importance (especially when the historical control data come from the same laboratory, using the same standardized diet, and do not go back in history beyond 5 years of the study under consideration) and that, if the chemical in question has only enhanced the incidence of a commonly-seen tumour to a level seen in historical controls, then the level of concern will be much less than would otherwise be the case.

The evaluation of studies in which these commonly-occurring tumours are a complicating factor need careful individual assessment. The tumours that have given rise to the most controversy in recent years are hepatomas (particularly in the mouse), pheochromocytomas in the rat (see below), lymphomas and lung adenomas in the mouse, pancreatic adenomas and gastric papillomas in the rat, and certain endocrine-associated tumours, including pituitary, mammary, and thyroid tumours, in both rats and mice. Some of these tumours, such as hepatomas and lung adenomas, may occur in the majority of untreated animals.

With the exception of lymphomas, some of which are virally associated, the endocrine-associated tumours, and possibly hepatomas in high-incidence strains of mice, which may involve oncogenes (47), there is no clue as to the origin of tumours that occur commonly in experimentally-used rodents. Indeed, there is not even any cogent speculation as to the mechanisms by which these tumour incidences are increased.

Adrenal medullary lesions in rats provide a good example of the problems encountered in interpreting the significance of high tumour incidences. An overview of the literature indicates that untreated rats of various strains may exhibit widely differing incidences of lesions described as "pheochromocytomas" (24, 48, 49). There are no clear criteria for distinguishing between prominent foci of hyperplasia and benign neoplasms, and pathologists differ in the criteria that they use for distinguishing between benign and malignant adrenal medullary tumours.

Rats fed *ad libitum* on highly nutritious diets tend to develop a wide variety of neoplasms, particularly of the endocrine glands, in much higher incidences than animals provided with enough food to meet their nutritional needs but not enough to render them obese. The adrenal medulla is just one of the sites affected by overfeeding. Controlled feeding, especially early in life, reduces the life-time expectation of developing either hyperplasia or neoplasia of the adrenal medulla in rats.

A complicating factor in assessing carcinogenicity studies is the question of how to consider benign tumours. If benign and malignant tumours are observed in an animal tissue and there is evidence of progression from the benign to the malignant state, then it is appropriate to combine the tumour types before performing statistical analysis. Assessment of the relative numbers of benign and malignant tumours at the various dose levels in the study can help determine the dose response of the animal to the compound under test. On the other hand, if only benign tumours are observed and there is no indication that they progress to malignancy, then, in most cases, it is not appropriate to consider the compound to be a frank carcinogen, under the conditions of the test (this finding may suggest further study). Often, how benign tumours should be considered is much less clear. Some clarification can be achieved by classifying and analysing tumours on the basis of their histogenic origin. This is helpful, not only for determining the significance of benign tumours, but also for preventing different malignant tumours occurring in the same organ from being grouped together for statistical analysis. For further discussion of these points, see (50), pp. 226-230.

The results of statistical analyses are often misunderstood and misused. An effect may be statistically significant but not be of any biological significance, because the animal's well-being is not affected by its occurrence. On the other hand, an event that is of biological significance, such as the occurrence of one or two tumours of a very rare type in treated animals, may not be significant by the usual battery of statistical tests. This difference between biological and statistical significance underscores the need for critical analysis of statistical results rather than the blind acceptance of the numbers obtained. The statistical aspects of the design and interpretation of toxicity studies are discussed in Annex II.

Generally, it is becoming increasingly difficult to identify a substance as a carcinogen with confidence. In particular, when animals with a high background incidence of tumours are involved, it is extremely difficult to know when to draw the line between a result that indicates that a compound that is potentially hazardous for man has been discovered compared with a compound that is merely an experimental curiosity. The International Agency for Research on Cancer (IARC) reviews evidence for the carcinogenicity of chemicals on a continuous basis and drafts monographs on many groups of substances. However, faced with the difficulty of separating different levels of "carcinogenicity", IARC working groups on the Evaluation of the Carcinogenic Risk of Chemicals to Humans designate compounds as being possessed of either "limited" or "sufficient" evidence for carcinogenicity. Given the natural desire of the food additive toxicologist to be as cautious as possible, this terminology is not very practical. After a compound has been designated as possessing "limited carcinogenicity", it is very difficult from a regulatory point of view to approve it as a food additive, even if extensive further work on the compound shows it to be safe at expected levels of consumption with no other evidence of carcinogenicity.

The decisions that are being made on the basis of the present state of knowledge of carcinogenesis, may, in the future, prove to have been excessively conservative. However, it is now possible to make certain reasoned decisions, provided that each instance where carcinogenicity is the problem is examined individually and all relevant factors are taken into account.

5.1.4 Reproduction/developmental toxicity

Most food additives are consumed by men and women during the reproductive stages of their lives and by pregnant and lactating women. Some food additives are also consumed by infants. Thus, a thorough safety evaluation requires that the effects of the substance on reproductive performance and development from fertilization through weaning be studied. JECFA recognizes, however, that it is unrealistic to expect that such studies be performed in all cases (sections 5.3.1 and 5.3.4).

Adverse effects on reproduction may be expressed through reduced fertility or sterility in either the parents or offspring due to morphological, biochemical, or physiological disturbances. Adverse effects on development may be expressed through structural or functional abnormalities due to either mutations or to biochemical or physiological disturbances. Mutations may occur in either somatic or germ cells. Mutations in male or female germ cells represent potentially the most long-lasting and severe effects on the human population that a chemical could cause.

Adverse effects on reproduction or development induced by chemicals may be expressed immediately or they may be delayed, sometimes for many years, as exemplified by transplacental carcinogens (section 5.3.1.2).

Structural or functional abnormalities are most likely to develop during embryogenesis, the period of development during which cells differentiate into the various organ systems. Typical teratogenicity studies investigate the effects of exposure to test substances during this period. Effects due to exposure during fetogenesis, the developmental period after the organ systems have formed, generally involve growth retardation and functional disorders, though the external genitalia and the central nervous system are also susceptible to injury during this period (51, 52). Such structural or functional abnormalities often do not become obvious until some time after birth and, in some cases, not until adulthood.

Neonatal development may be influenced by the consumption of milk containing chemicals (or their metabolites) that were consumed by the mother. Agents may also affect neonatal development by influencing maternal behaviour, hormonal balance, or nutrition. Direct neonatal exposure to xenobiotic compounds also occurs, but is less common, since JECFA considers it prudent that food intended for infants younger than 12 weeks of age should not contain any additives (42).

Guidelines for reproductive toxicity investigations have been developed by various legislative and international organizations, including the US Food and Drug Administration (US FDA), US Environmental Protection Agency (US EPA), the United Kingdom Committee on Safety of Medicines (CSM), Committee on Toxicity (COT), and Pesticides Safety Precautions Scheme (PSPS), the Japan Ministry of Agriculture, Forestry and Fisheries (MAFF) and Ministry of Health and Welfare (MHW), the World Health Organization (WHO), the Organization for Economic Cooperation and Development (OECD) (all listed in (51)) and the International Programme on Chemical Safety (IPCS) (53). A review of the methodology for assessing the effects of chemicals on reproductive function has been published under the auspices of the IPCS and the Scientific Committee on Problems of the Environment (SCOPE) of the International Council of Scientific Unions (54). The procedures described in these publications are designed to assess the reproductive and developmental toxicity potential of test compounds using lower mammals as model systems. These procedures generally involve combining various stages of the life cycle in one test, as it is usually not practicable to examine the effects of a chemical in each separate stage of the reproductive cycle. An exception is the so-called "teratogenicity" study, where exposure is limited to the period of organogenesis (see below).

The goal of reproduction/developmental toxicity studies is to assess whether the organism is more sensitive to the agent under test during its reproductive and developmental stages than during its adult phase. Therefore, the highest dose of food chemical that is administered is generally the amount that would be expected to cause slight maternal toxicity, and the lowest dose is the amount that is not expected to cause an effect in either the mother or the conceptus. If profound toxicity is observed in the offspring at the high dose (the dose that causes only slight maternal toxicity), then the conclusion would be that the substance is more toxic for offspring than for adults. This conclusion would be reinforced by the appearance of adverse effects in the conceptus at the mid- and/or low-dose levels. On the other hand, if the test substance injures reproduction or development at levels comparable with levels that cause toxicity in adults, then no special concern should be attached to the results of the reproduction/developmental toxicity studies.

Single-generation and multi-generation reproduction studies are useful for assessing potentially deleterious effects on reproduction and development through parturition and lactation. However, because of the long-term exposure inherent in these studies, detoxifying enzymes may be induced in mothers before embryogenesis takes place. Under these circumstances, the observed toxicity would be understated. Studies in which mothers are exposed to the test substance only during organogenesis, as in teratogenicity studies, reduce the possibility of the mother adapting to the test compound.

The range of effects arising from maternal exposure to chemicals during organogenesis includes:

(a) death and resorption of the embryo;

(b) teratogenic defects (malformations of a structural nature);

(c) growth retardation or specific developmental delays; and

(d) decreased postnatal functional capabilities (55).

Which of these effects will be expressed depends on the level and gestational timing of the dosage of the food chemical, and the duration of the period of treatment (56). Thus, a substance given at one dose level may result in growth retardation, while, at a higher level, it may result in death and resorption of the embryo. Sometimes, the slope of the dose-response curve of, and between, these effects is very steep, making the interpretation of the studies very difficult. Because all of these outcomes are unacceptable, the most important consideration when evaluating these studies should not be which effect is observed, but rather, at what dose level the adverse effect became evident. This dosage information can then be used to set exposure limits. Because teratogenic effects are only one part of the total spectrum of the embryotoxic effects that should be investigated in such studies, a better term for "teratogenic studies" might be "embryotoxicity studies". In those rare situations when the studies are performed during the period of fetal development, the term "fetotoxicity studies" should be used.

The appropriate role of studies involving *in utero* and neonatal exposure in the evaluation of food additives is discussed in section 5.3.

5.1.5 *In vitro* studies

In recent years, a great deal of effort has gone into the development of *in vitro* test systems. Generally, these systems are segregated according to two kinds of functions: (a) to reveal whether a particular kind of toxicity is produced by the agent under study; or (b) to help elucidate the mechanism of toxicity displayed by a chemical. The former tests are being developed to serve both as predictors of toxicity (section 3.1.2) and as substitutes for complex, lengthy *in vivo* procedures. The latter are more directly focused than the former, and their value has been clearly demonstrated as a means of establishing the metabolic mechanisms at the organ, tissue, or cellular level (section 5.2).

Much effort has been devoted to the development of *in vitro* test systems based on isolated cells, tissues, and organs. Some of these systems are reported to be related to specific toxic end-points such as mutagenicity and carcinogenicity (e.g., DNA damage and repair in mammalian cells, covalent binding to DNA, cell transformation, mitotic recombination, and gene conversion in yeast (57)) and to embryotoxicity (e.g., whole embryo cultures, cultures of embryonic tissues, teratocarcinoma cells, and embyronated eggs (58)).

The number, diversity, and use of these tests have increased rapidly in the past decade and are likely to continue to increase in the future. However, correlations among results of various *in vitro* tests and reported correlations between the results of batteries of short-term *in vitro* tests and *in vivo* carcinogenesis bioassays (which have been the primary thrust of these assays) are not high. Such short-term *in vitro* tests are generally effective at measuring their intended genetic endpoint, i.e., mutagenicity in the particular system under study. However, the relevance of mutagenic effects to food additive toxicity has not been established, and the results of many current *in vitro* test procedures do not relate to genetic effects in mammalian reproductive tissues. Neither is it clear how well these tests identify chemical carcinogens or how they should be used in the absence of corroborative data on carcinogenicity.

In a similar fashion, culture techniques designed to measure prenatal toxicity are extremely useful for research purposes, but, at their present stage of development, they are not very suitable for screening (58). By excluding the maternal-placental-fetal relationship, such dissected systems permit the compound to reach the target directly (membrane systems that provide biological barriers are missing) without permitting the potential moderating or activating influences of the maternal tissues.

Attention should be paid to scientific developments in *in vitro* test systems. However, because of the many experimental uncertainties and controversial issues surrounding the efficacy of these tests as predictors of specific toxic end-points, it would be inappropriate for JECFA to request that all food additives brought before it should be subjected to such tests on a systematic basis. On the other hand, data obtained with *in vitro* systems sometimes help to clarify the mechanism of action of chemicals observed in *in vivo* systems. Therefore, JECFA should continue to determine the relevance of available *in vitro* data on an *ad hoc* basis when assessing the safety of specific compounds.

5.2 The Use of Metabolic and Pharmacokinetic Studies in Safety Assessment

Chemical toxicity results from reactions between the ingested toxic chemical, or its metabolites, with constituents of the body. Therefore, the complete safety evaluation of a substance such as a food additive must consider its metabolism and pharmacokinetics. Unfortunately, a great deal more has been said and recommended in this area than has been done in practice. The importance of metabolic and pharmacokinetic data in the proper planning and interpretation of *in vivo* toxicity testing of chemicals is obvious, but the fact is that such data are either inadequate or unavailable to aid in the interpretation of the majority of long-term studies undertaken with chemicals, including food additives.

Detailed metabolic studies have gained added importance in determining the extent of appropriate toxicological testing since the advent of novel and modified foods. This is considered in detail in section 6.2. However, it is necessary to repeat some general aspects of the subject here.

Biochemical studies play two separate roles in the safety evaluation of chemicals. These are:

(a) to design animal studies by identifying the appropriate species for, and helping to determine the appropriate level of, testing; and

(b) to extrapolate experimental animal toxicity data to human beings, by elucidating the mechanism of toxicity of the chemical, thus facilitating the establishment of a no-observed-effect level; a comparison of biochemical data between experimental animals and man helps determine the relevance of any toxicity observed in animals.

The ingested chemical itself may exert a toxic effect, or a metabolite(s) may be the toxic agent. Many polar, non-lipophilic chemicals are rapidly metabolized and/or excreted, while lipophilic compounds may be stored, excreted into the bile, or metabolized into more polar, water-soluble compounds, which are eliminated from the body, in the urine, more rapidly than the ingested additive.

Absorbed substances, except those that enter the lymphatic system, are transported directly to the liver via the portal vein. Many substances that are metabolized in the liver are transported via the hepatic vein to the kidneys to be excreted in the urine. Through enterohepatic circulation, some substances that are conjugated in the liver are excreted with the bile, reabsorbed, and then excreted once again in either the bile or the urine.

Metabolism, primarily involving enzymatic reactions, may:

(a) convert the additive into a body constituent or a source of energy;

(b) lead to the detoxification of the ingested chemical and the excretion of its metabolites; or

(c) result in activation of the chemical into reactive intermediates that then react most importantly with glutathione, tissue proteins, RNA, or DNA.

Biotransformation reactions are catalysed by intra- and extracellular enzymes and by enzymes of the microflora of the gastrointestinal tract. Knowledge of the rates of formation, reaction with tissue components, and excretion of various metabolites is essential for full understanding of the disposition and elimination of the chemical from the body and of the mechanism and extent of its toxicity.

This section contains a general discussion of the role of metabolism and pharmacokinetic data in the safety assessment of food additives. Simple guidelines have not been generated, as it is doubtful that such guidelines are feasible or desirable. Several food additives that have been extensively studied biochemically are discussed in Annex IV. These examples are designed to provide an appreciation of the value and problems involved with investigating the metabolic bases for the biological responses to food additives.

5.2.1 Identifying relevant animal species

The occurrence of interspecies differences in response to foreign compounds complicates the extrapolation of animal toxicity data to human beings. The resolution of this problem depends on an understanding of such interspecies variations in the disposition of ingested compounds. In this context, disposition is meant to encompass metabolism and pharmacokinetics.

The rates of absorption, rates and sites of distribution, and rates and routes of excretion determine the concentration-time profiles of the parent molecule and metabolites in the various tissues and organs of the body. The overall biological response is thus the product of the fluxes of the unchanged molecule and its metabolites occurring in the animal under examination. Definition of the pharmacokinetic properties of a food additive may require various routes of administration. The influence of any vehicle to be used in long-term studies should be determined, because the vehicle may influence the absorption, metabolism, or toxicity of the test compound.

In order to extrapolate reliably from animals to man, the ideal situation would be one in which the tissues of the animals and of man would be exposed to identical fluxes of the compound and its metabolites. This requires that the qualitative, quantitative, and kinetic aspects of the disposition of the compound be the same in animals and man. This ideal situation is probably never achieved because of species variations.

The goal should be to select a species for testing that is the most closely related to man in terms of the metabolism of the compound under study, using a route of administration similar to the anticipated human exposure. However, the list of species that may realistically be used in a toxicity test is very limited because of problems of availability, lack of background pathological and physiological knowledge, and experimental convenience. Thus, it is unlikely that a suitable metabolic model species will fulfill other important criteria used to select test species. Given these facts, metabolic studies used for species selection should be prospectively performed only on species suitable for toxicity testing. The species is then selected that is closest to the human being in terms of the metabolism of the compound. This, of course, requires knowledge about its metabolism in the human being. In many cases, *in vivo* human studies are not possible.

In general, the required metabolic and kinetic information can best be obtained from *in vivo* studies. Pharmacokinetics can obviously only be examined *in vivo*, since these studies deal with whole animal phenomena. *In vitro* studies, such as organ perfusions and tissue cell incubations, provide useful information in some cases, but they do not provide information on the absorption, distribution, and excretion of chemicals. The use of isolated cells of human tissues may prove acceptable in some cases, because, even considering their inherent limitations, these systems may be the only ones available for examining the metabolism of compounds that cannot be administered to human subjects.

In a long-term toxicity test, the attainment of a steady state depends on the relationship between the kinetic variables of the compound and the dose interval. When a compound is administered continuously in the diet, an approximate "steady state" will sometimes be established. Therefore, marked differences seen in animals in single-dose pharmacokinetic studies may change, or even disappear with long-term administration. However, a steady state will never truly be achieved because of diurnal patterns of dietary intake by common laboratory animals.

On the other hand, compounds that are rapidly absorbed and have a short half-life will show wide temporal variations in plasma concentrations, as will compounds that are administered by gavage or capsule.

In many cases, the metabolic profile of a compound is determined by the amount administered, as well as the species in question. The use of very high doses in toxicity testing may give metabolic patterns, and therefore biological responses, that are patently unrepresentative of the situation to be expected with actual levels of exposure. Thus, data on the influence of dose level on metabolism in the test animal should be generated to determine whether absorptive, metabolic, or excretory processes may have a threshold. An animal model, apparently suitable at one level of exposure, may be less suitable at a different level.

Consideration should be given to the possibility that the metabolism of the compound will differ between long-term tests and short-term metabolic studies. This could be because of adaptation by gut microflora, which is discussed in section 5.2.4, or because of the induction of enzyme systems that metabolize the substance.

An aim of toxicity testing should be to examine the possible activities in animals of all of the human metabolites of a compound that may induce toxicity. In many cases, this is best achieved by combining data from several animal species, to include all the metabolites of interest. In interpreting such data, the closest attention should be paid to the similarities of the mechanisms of toxicity in the various animal species, and also to the possibility that toxicity may involve interactions between the parent compound and its metabolite(s), which may not be the same in all species and may be irrelevant for man.

5.2.2 Determining the mechanisms of toxicity

A great deal of research has been performed in an attempt to explain the mechanism(s) by which certain test chemicals have given rise to particular lesions. In the majority of cases, these studies have concerned the development of tumours. Clearly, it is likely to be most difficult to find any simple mechanistic answer to carcinogenesis. However, it is usually possible to determine factors of importance for a safety evaluation, which are not necessarily complete solutions to the mechanism of action.

If an additive has been determined to be an animal carcinogen, it is extremely difficult to show that it is safe for human beings. It is not easy to establish its safety, with retrospective metabolic and pharmacokinetic studies, that an exclusively secondary mechanism is operative and that a threshold exists below which the use of the additive is safe.

If a carcinogenic impurity is present in an additive, the impurity should be either removed or limited to such an extent that consumption of the additive does not pose a carcinogenic risk for consumers. The level of the impurity in the food additive should be controlled within specifications.

In some cases, toxicity may occur as a result of the test compound or a metabolite displacing endogenous substrates from carrier proteins or receptor sites. Where such mechanisms are indicated, *ad hoc* studies of relative binding affinities can form a useful adjunct to routine pharmacokinetic studies and assist in establishing safe levels of exposure. *In vitro* studies are useful for determining the mechanism of toxicity, especially when covalent binding to cellular macromolecules is involved. In such cases, comparative binding studies using preparations of metabolizing enzymes from various animal species are desirable.

5.2.3 *Metabolism into normal body constituents*

The metabolism of an ingested compound into normal body constituents does not provide assurance that the substance is safe. Not only are many metabolites toxic (e.g., most excretory products), but there are limits to the body's ability to process even relatively non-toxic metabolites. These limits should be known, and, if a toxic threshold for a substance has been identified, its acceptance as a food additive will depend on controlling intake so that toxic levels are not ingested by human beings.

Knowledge that a substance is a natural metabolite or is metabolized into normal body constituents is of great help in evaluating its safety. However, without concomitant information on the kinetics of the production and disappearance of metabolites, the extent to which such information can be used is limited.

When the additive provides only a small increment in levels of metabolites compared with the ordinary consumption of food, then the questions about safety are greatly simplified. The report of the WHO Scientific Group on Procedures for Investigating Intentional and Unintentional Food Additives (2, p. 7) considered this situation and concluded that:

> "if the biochemical evidence shows that the additive makes only a small contribution to existing metabolic pools from food components or in the tissues, there may be no need for detailed toxicological studies on it, provided that it conforms to adequate specifications."

JECFA has also considered this situation in some detail (29, pp.12-13) and has suggested that:

> "any food additive that is completely broken down in the food or in the gastrointestinal tract to substances

that are common dietary or body constituents might be satisfactorily evaluated. . . on the basis of appropriate biochemical and metabolic studies alone. . .".
This report summarizes the evidence required in such cases as follows:

(a) "evidence that the substance is readily broken down in the food or in the gastrointestinal tract to common food constituents under the conditions of use;

(b) evidence to indicate the main factors concerned in this breakdown, e.g, pH and enzymes;

(c) evidence, preferably including studies on human subjects, that the material, when given in moderate amounts and under conditions similar to those that will prevail if used as a food additive, is absorbed to the same extent as the food materials to which it gives rise, and does not interfere with the absorption of other nutrients;

(d) evidence that unhydrolysed or partly hydrolysed material does not occur in significant amounts in the stools, and that it does not cumulate in body tissues; and

(e) evidence that the most important food components in the additive are metabolized and utilized as effectively when administered in composite form as when given separately, and that overloading does not occur."

As long as adequate evidence along the above lines is presented, the Committee concluded that:
"the food additive is handled in the body in a way that is not significantly different from that required for the component food materials. If so, no toxicological studies need be demanded, since the problem now becomes one involving the toxicology of foods themselves rather than the toxicology of a food additive."
The Committee allocated ADIs to such substances, calculated on the basis that the food additive would not increase the food component into which it is converted by more than about 5% of the quantity in an average diet (29, p. 13).

These general principles have been accepted by the eleventh and seventeenth meetings of JECFA (41, pp. 8-9; 16, p. 31), the second of which confirmed that, if biochemical evidence shows that the sole effect of the additive is to make a small contribution to existing metabolic loads from food components, there is no need for detailed toxicological studies. Examples cited in the various reports include sucrose esters of fatty acids, lactic and fatty acid ester of glycerol, and some esters used as food flavours.

These principles are still valid. However, in some JECFA reports, the combined evidence of breakdown in food and in the gastrointestinal tract was considered. In contrast, studies on the stability/breakdown pathways of the additive in food, under the proposed conditions of use, may be needed to ensure that significant quantities of toxic products are not formed during food processing or storage, either through transformation of the additive or through its reaction with food constituents. All procedures designed to measure metabolites must be accurate and they must have a high level of sensitivity for the compounds under consideration, to draw the conclusion that further toxicity studies are not necessary.

5.2.4 Influence of the gut microflora in safety assessment

The gut microflora may influence the outcome of toxicity tests in a number of ways, reflecting their importance in relation to the nutritional status of the host animal, to the metabolism of xenobiotics prior to absorption, and to the hydrolysis of biliary conjugation products. JECFA has recognized this, and has drawn attention to the usefulness of studies on metabolism, involving the intestinal microflora, in toxicological evaluation (30, p. 7; 18, p. 10).

Interactions that may occur between food additives and the bacterial flora of the gastrointestinal tract should be considered both in terms of the effects of the gut microflora on the chemical and the effects of the chemical on the gut microflora. Because the gut microflora are important in the metabolic fate and toxicological activity of some food additives, the safety assessment of food additives should include the possibility that gut microflora modify the host response to the food additive and/or that the food additive is affecting the host microflora.

5.2.4.1 Effects of the gut microflora on the chemical

The spectrum of metabolic activity shown by the gut flora contrasts markedly with that of the host tissues. While hepatic metabolism of foreign compounds is predominantly by oxidation and conjugation reactions, the gut bacteria perform largely reductive and hydrolytic reactions, some of which appear to be unique to the gut flora. Typical reactions include:

(a) the hydrolysis of glycosides (including glucuronide conjugates), amides, sulfates, and sulfamates;

(b) the reduction of double bonds and functional groups; and

(c) the removal of functional groups such as phenol and carboxylic acid moieties.

Thus, from a structural point of view, many food additives are potential substrates for microbial metabolism.

The gut bacteria are situated principally in the terminal parts of the intestinal tract, and thus highly lipid-soluble compounds that are absorbed in the upper intestine will not undergo bacterial metabolism. However, tissue metabolism may give rise to conjugates that are excreted into the bile and thus available for bacterial hydrolysis. Clearly, then, the design of appropriate investigations with the gut microflora must be linked closely to *in vivo* studies on absorption and metabolism. *In vitro* incubation of the food additive and/or its metabolites with the bacteria of the caecum or faeces is a useful but difficult technique, with considerable potential for the generation of spurious data. Some of the pitfalls of prolonged incubations are that:

(a) the use of a nutrient medium may allow the growth of a non-representative bacterial population; while

(b) the use of a non-nutrient medium may act as a powerful selective force for organisms able to use the additive as a source of carbon and energy.

There are three primary *in vivo* methods for studying the role of the gut microflora in the metabolism of a compound:

(a) parenteral administration of the compound, which should result in decreased microbial metabolism of poorly absorbed polar compounds, compared with oral dosing;

(b) studies on animals in which the bacterial flora are reduced by the use of antibiotics or by surgical removal of the caecum; and

(c) studies on germ-free animals and on (formerly) germ-free animals contaminated with known strains of bacteria.

A number of factors may influence the metabolic activation of foreign chemicals by the host microflora (see reference 59 for an expansion of these points):

(a) Host species

Species differences exist in the number and type of bacteria found in the gut and in their distribution along the gut. In this respect, the rat is a poor model for man, since significant numbers of bacteria occur in the upper intestinal tract of the rat, whereas this region is almost sterile in man.

(b) Individual variations

There is a great deal of variability among individuals within a species in the extent to which some compounds undergo metabolism by the gut flora. Many of these variations probably arise from differences in the enzymatic capacity of the gut flora rather than in the delivery of the chemical to the lower intestine. Thus, if, in animal studies, a food additive is shown to be metabolized by the gut flora to an entity of toxicological significance, it is essential that its metabolic fate be characterized in the human being.

(c) Diet

The composition of the gut flora depends on the diet, which may influence the extent of microbial metabolism of a food additive.

(d) Medication

The widespread oral administration of medications, such as antibiotics and antacids, in the human population, is a cause of variations in metabolism by the gut microflora.

(e) Metabolic adaptation

The metabolic capacity of the gut flora is far more flexible than that of the host. Thus, long-term adminstration of foreign chemicals can lead to changes in both the pattern and extent of microbial metabolism of the chemical. Because prior exposure to the compound under test may significantly alter the metabolic potential of the gut microflora, metabolic studies should be performed not only on previously unexposed animals but also on animals that have been exposed to the test compound for some time. For the same reason, any *in vitro* studies should be performed with caecal contents that have been collected both prior to and during long-term animal feeding studies.

5.2.4.2 Effects of the chemical on the gut microflora

During high-dose animal feeding studies, the gut microflora may be affected in two ways:

(a) Development of antibacterial activity

A weak antibacterial activity may become significant after long-term intake of near-toxic doses of a food additive. This may manifest itself either as an alteration in the number of bacteria present, which can be measured directly, or as an abnormal microbial metabolic pattern. The latter can be studied by measurement of certain endogenous metabolites produced only by the gut flora, such as phenol and p-cresol, which provide indirect evidence of alterations in the gut flora. Such information may also be of value in the interpretation of other variables such as nitrogen balance.

(b) Increased substrate for gut microflora

The food additive may act directly as a substrate for bacterial growth. This can be readily illustrated by appropriate high-dose pharmacokinetic studies, coupled with *in vitro* metabolic studies on the gut flora. Alternatively, the food additive may inhibit digestion or absorption of other dietary components so that these become available to the bacteria in the lower intestine in increased amounts.

Increased amounts of substrates in the lower intestine provide an increased osmotic effect in the caecum, which may be detectable as caecal enlargement (section 5.1.2). The reason for caecal enlargement must be studied before the significance of the lesion can be assessed since it may indicative of:

(i) abnormal osmotic balance with consequent changes in permeability to minerals in the caecum, which could lead to nephrocalcinosis;

(ii) microbial metabolism of nutrients, which could result in the formation of potentially toxic metabolites and abnormalities in the nitrogen balance; or

(iii) microbial metabolism of the food additive, which could lead to the formation of toxic products.

5.3 Influence of Age, Nutritional Status, and Health Status in the Design and Interpretation of Studies

Animal toxicity studies are generally performed with healthy animal populations that are in a state of over-nutrition in a protected environment. This basic procedure is altered only when there is a specific reason for doing so, for example, when nutritional factors are being studied.

In order to establish the safety of food additives, experimental protocols have tended towards more universal designs encompassing populations during all stages of the life cycle, e.g., reproduction studies are often included in long-term studies. Such protocols are intended to mimic the type of exposure in the bulk of the human population. Margins of safety and medical advice are used to protect subpopulations at special risk for one reason or another.

5.3.1   Age

It is well known that the age of a test animal can influence the toxic response to a substance being tested. For example, an enzyme activity that is involved in the metabolism of a substance in an adult may be virtually absent in an immature animal or *vice versa*. Thus, a compound that is metabolized to a less toxic metabolite in an adult animal would be more toxic for young animals lacking the appropriate enzyme activity; obviously, the reverse would be true for a substance metabolized to a more toxic metabolite. Differences in sensitivity between mature and young animals may also result from differences in intestinal flora as observed, for example, in the growth of distinctive flora in the upper intestine in human infants that render them sensitive to nitrate. Greater sensitivity may also arise in young animals, because of the incomplete formation of intestinal, blood-brain, or other tissue barriers, which leads to the passage of potentially harmful substances through the barriers.

5.3.1.1 History

The WHO Scientific Group on Procedures for Investigating Intentional and Unintentional Food Additives discussed the effects of age on toxicity (2, pp. 10-12) and found that "in general, the young animal is more sensitive to the toxic effects of exposure to chemicals". Among the reasons cited for the increased sensitivity in neonates were differences in the distinctive flora of the upper bowel and differences in the levels of the "drug-metabolizing enzymes", which are frequently low in the neonate. Attention was drawn to interspecies differences in the neonatal levels and in age-related changes in the levels of these enzymes. The Scientific Group stated that "pertinent information derived from reproduction (multi-generation) studies provides some assurance on the safety of compounds that might be present in the diet of babies" but felt that "since babies constitute a special population, close observation of epidemiology in this group is an important practical aspect of the evaluation of the effects of exposure." The Scientific Group also saw the need for "further information on the development of enzyme systems in the human young, with particular emphasis on those enzymes responsible for dealing with foreign compounds." The report concluded (2, p. 23) that "useful information may be obtained from studies in newborn or young animals, from reproduction studies, and from biochemical studies" and called for further research on "the development of enzyme systems in the human young, with particular emphasis on those enzymes responsible for dealing with foreign chemicals" (2, p. 25). With respect to the latter research, the Scientific Group concluded that "this information is essential in assessing the safety of additives in baby food."

Subsequently, the tenth Report of JECFA (29, p. 24) recommended that a special subcommittee of JECFA should be established to study the special problems arising from exposure of infants and young children to food additives. In response to this recommendation, an FAO/WHO meeting on Additives in Baby Foods was convened in 1971, and the report of this meeting was included as Annex 3 to the fifteenth JECFA Report (42, pp. 29-37). A distinction was made, on developmental grounds, between children up to 12 weeks of age and children over 12 weeks. The Subcommittee considered it prudent that food intended for infants under 12 weeks of age should not contain any additives at all. However, if it were deemed necessary to use additives in food intended for young infants, the Subcommittee concluded that "particularly for infants under 12 weeks, toxicological investigations should be more extensive and include evidence of safety to young animals."

With respect to contaminants, the Subcommittee concluded that "the establishment of acceptable residue levels of pesticides or other contaminants likely to be present in milk and cereals for infant foods should be based on toxicological evaluation in very young animals" (42, p. 31). The report also made observations on particular classes of food additives (42, pp. 32-33). The vulnerability of very young infants was recognized, and guidelines on toxicological testing were formulated (42, p. 34). These include the following:

(a) Before a food additive is regarded as safe for use in food intended for infants up to 12 weeks of age, the toxicological studies should be extended to include animals in the corresponding period of life.

(b) It is difficult to recommend precise toxicological testing procedures until more basic research has been undertaken. There are also difficulties in selecting appropriate species. In these circumstances, short-term studies should be conducted in several species and should include the oral administration of the additive under test, at suitable dose levels, to newly born animals up to and including the end of the weaning period.

(c) When life-span studies and multi-generation studies are carried out, they should be extended to include oral adminstration of the food additive at suitable dose levels to a proportion of animals from the day of birth throughout the pre-weaning period.

The practical difficulties and cost of implementing these recommendations on a routine basis would be immense, involving, as it would, artificial feeding of litters of newborn laboratory animals. However, in situations in which young infants are a target population for an additive, it seems reasonable that studies such as these should be performed.

When considering glutamate, the fourteenth report of JECFA (30, p. 8) noted that:

"any attempt to interpret these data in terms of human neonates and infants involves the problem of how far developmental stages in animal species and in man can be considered equivalent in relation to vulnerability to possible effects of food additives. Relevant information would be of considerable value."

The sixteenth (60) and twentieth meetings (19, p. 22) of JECFA recommended that a review should be made of the special problems arising from the exposure of infants and children to contaminants in food. This review was conducted at the twenty-first meeting (20, pp. 9-12), which also considered food additives. The Committee stated that:

"scientific evidence indicates that newborn and very young children are particularly sensitive to the harmful effects of foreign chemicals" due to, *inter alia*, "immaturity of enzymatic detoxifying mechanisms, incomplete function of excretory organs, low levels of plasma proteins capable of binding toxic chemicals, and incomplete development of physiological barriers such as the blood-brain barrier. Moreover, there appears to be a general vulnerability of rapidly growing tissues, which is particularly important with regard to the developing nervous system."

The Committee reiterated that food intended for infants under 12 weeks of age should (with certain exceptions) not contain any additives but that "in assessing food additive safety, the question of potential special hazards for the newborn and infants should be kept in mind" and "toxicological and metabolic studies of food additives should always include investigations that permit the evaluation of safety for the newborn and the infant." The Committee stated further that "in order to gain more information about the long-term effects of exposure in utero and in the post-natal period, appropriate methodology must be developed," and the Committee emphasized that "short- and long-term effects of exposure in utero and during lactation should be taken into account for food additives and contaminants evaluated by the Committee" and "this evaluation might include a request for appropriate animal studies."

Implicit in these statements is a call for metabolic studies on neonates and for toxicological studies involving in utero exposure followed by long-term studies. Indeed, the twenty-second report of JECFA (32, p. 30) reaffirmed the need for testing the effects of exposure to food additives and contaminants in utero and on neonates during suckling. However,

"in view of the complexity of the testing procedures," the Committee recommended that "WHO should convene a meeting of experts to assess: (a) the degree of any increase in the sensitivity of toxicological testing afforded by exposure in utero through lactation; and (b) the need to include such exposure in toxicological tests as a means of increasing public health protection."

Criteria for determining whether in utero exposure is to be included in such studies should include such information that the chemical crosses the placental barrier and/or is secreted in breast milk. The Committee recommended further that:

"the experts should also propose the most appropriate guidelines for experimentation, taking into account: (a) the dosages used and the relative exposure of mother and fetus to the agent under study; (b) the possibility of combining this modified long-term test with reproduction studies; (c) the length of the studies required; and (d) the most appropriate species to use."

A meeting of experts has not yet been convened to consider these issues.

A document of potential benefit to JECFA is one that has recently been developed by the IPCS and the Commission of the European Communities concerning the principles for evaluating health risks from chemicals during infancy and childhood. Among the objectives of this activity were to:

(a) "investigate whether and when there is a need for specific approaches when evaluating the health risk associated with exposure to chemicals . . . during infancy and childhood"; and (b) "identify further developments in methodology that are necessary for the assessment of health risks associated with exposure to chemicals during the early period of life" (61).

5.3.1.2 Usefulness of studies involving *in utero* exposure

In order to evaluate the usefulness of *in utero* studies, it is important to review the available toxicity data relating to this issue. Most studies including *in utero* exposure have involved the use of known carcinogens. The IARC monograph "Transplacental Carcinogenesis" (62) provides information on much of the earlier research on these substances. In general, the reports indicate that, although transplacental carcinogenesis could occur, there have been no instances in which compounds that were carcinogenic for the offspring were not also carcinogenic for the adult and *vice versa*. However,:

(a) the carcinogenic effects in the offspring can occur at sites different from those observed in the parent (e.g., in transplacental rat studies, ethylnitrosourea shows striking neuro-oncoselectivity, not observed in the parent, and the induction of unique tumours of the vagina in young women has been observed with diethylstilboestrol treatment of mothers given the drug for pregnancy maintenance); and

(b) the fetus may be more susceptible to tumour development than the adult rat (observed with ethylnitrosourea).

On the other hand, in some cases, adults are more sensitive than young offspring, as with certain nitrosamines (62, 63), suggesting that enzyme systems capable of converting these compounds into their ultimately carcinogenic forms are not fully developed in fetal tissue.

Although transplacental carcinogenesis has been the major interest of *in utero* studies, evidence is accumulating that non-carcinogenic substances may be the cause of a variety of biochemical and other toxic effects in the developing fetus. Some of this information has come from studies of the toxic effects of environmental contaminants such as methylmercury and PCBs. Poisoning episodes of methylmercury in Japan and Iraq indicate that the developing fetus shows toxic symptoms at levels at which the mother is asymptomatic. This appears to be because of selective localization of methylmercury in the brains of exposed fetuses rather than a higher sensitivity of the fetus itself (64). In the case of PCBs, monkeys with body burdens of PCBs derived from previous exposure produced offspring that showed behavioural and learning deficiencies; it was estimated that approximately 40% of the body burden of PCBs in the offspring was derived from placental transfer (65). A finding of equal or greater interest in this study was that 60% of the body burden of PCBs was transferred postnatally in the milk. This result is consistent with the finding that the only route of excretion of certain chemicals from the human body, particularly the halogenated hydrocarbons that accumulate in fat, may be via breast milk. Breast-fed infants, therefore, may be exposed to very high levels of these compounds, far exceeding the acceptable daily intake (ADI) or the provisional tolerable weekly intake (PTWI) (66), pointing to the critical need for obtaining data during the neonatal phase for these types of compounds.

The results of transplacental rat studies have also shown the possibility that the course of enzyme development in the fetus may be markedly altered by exposure to foreign substances. This so-called programming may alter the time of development of specific enzymes or change the pattern of development of sex-dependent enzymes, i.e., male offspring may develop enzyme profiles more characteristic of female offspring (67). Studies designed so that the progeny of exposed parents are used in the long-term phase serve as a fitness test to detect subtle effects of this type.

Possible differences in the placenta structure in human beings and experimental animals should be considered in the interpretation of *in utero* studies. Structural differences may result in significantly different rates of transfer of chemicals across the placental membrane in experimental animals compared with human beings. This should be considered when selecting appropriate dose levels during the *in utero* phase of animal studies.

5.3.1.3 Complications of aging

The Scientific Group on Procedures for Investigating Intentional and Unintentional Food Additives (2) concluded that it is "better to carry out toxicity studies before the complications of senescence arise" but, nevertheless, called for "more basic information . . . on toxicity in aged as well as in young animals."

Older animals may be especially sensitive to certain substances, because of reduced functioning of vital organs such as the kidney or liver. The ability to metabolize certain substances in the liver may decrease in aged animals (68-71), resulting in an accumulation of toxic substances and consequent effects that would not be seen in young animals. Normally-occurring lesions of old age such as tumours or kidney lesions may mask subtle compound-related pathology and may render aged animals a poor model system for assessing some lesions. Conversely, some lesions may require long exposure to develop or may only be manifested in older animals. Such old animals are used routinely, and more research and documentation are needed in this area.

5.3.2 Nutritional status

JECFA has not directly addressed the issue of over-nutrition in laboratory animals. Excessive food intake and, in particular, *ad libitum* feeding of animals can complicate the interpretation of studies. Considerable research indicates that altered caloric intake and qualitative changes in diet can have a profound effect on various disease processes, particularly on the occurrence of neoplasms (72, 73). Additional research is needed leading to better nutritional designs of experimental models for safety evaluation.

In considering the effects of nutritional status on toxicity, the WHO Scientific Group on Procedures for Investigating Intentional and Unintentional Food Additives (2, pp. 12-13) recognized that nutritional status can influence toxicity, positively or negatively, depending on the substance, but felt that "it is wise to maintain all the animals on a diet that is nutritionally adequate in every way, unless there is some specific reason for doing otherwise." While noting that "further work is needed on the effects of various states of malnutrition or undernutrition on the toxicity manifested by chemical compounds," the Scientific Group concluded that "an effort to simulate conditions of malnutrition in man . . . is not considered advisable in routine toxicological investigations intended for the evaluation of safety" and "the evaluation of safety is best carried out by using healthy animals on adequate, balanced diets."

In the seventeenth JECFA report (16, p. 31), it is noted that reactions of food additives with food constituents may affect the nutritional value of the food and that this possibility can be studied by chemical or biological assay methods (section 4.2). However, it is further reported that "it may be necessary to undertake a toxicological investigation of treated food materials; here, a margin of safety may be introduced by conducting the test with food that has been deliberately over-treated to a measured extent." In addressing this problem of interaction between additives and food constituents in the twenty-fourth report of JECFA (21, pp. 10-11), it is pointed out that such reactions may occur during food manufacture, storage, and cooking, and it is re-emphasized

"that a better perspective of the safety of food additives would be gained if information on their manufacture and technological use were more readily available. Such information should cover . . . any available data on the chemical fate of each additive in those foods and on the effects of additives on nutrients . . . ."
It "may even be necessary sometimes to carry out a study on the technological versus nutritional effects of certain additives and to present this information to the Committee."

Similar conclusions were reached by the twenty-fifth meeting of JECFA (22, pp. 11-12).

In the twenty-fourth report of JECFA (21, p. 10), increasing concern is also expressed with "the development of materials designed as substitutes for normal components of food" and noted that "questions of nutritional adequacy arise in such instances and must not be overlooked." The Committee believed that "the problems associated with the designing of tests to assess toxicity of these substances and with their interpretation and extrapolation to man require special consideration".

In considering the particular case of acceptable daily intakes (ADIs) with regard to nutrients such as ascorbic acid used as food additives, it is noted in the eighteenth report of JECFA that the lower limits corresponding to the requirements for such nutrients are determined by expert committees "concerned with adjusting the ADI if a food additive is shown to interfere with nutritional requirements in one form or another" (17, footnote p. 9).

Interference with nutritional requirements can occur by antagonizing the normal physiological function of a vitamin, trace metal, or other micronutrient, either through destruction of the micronutrient before ingestion (such as thiamine destruction by sulfur dioxide) or through antagonism or inactivation in the body after ingestion. Testing regimes with both nutritionally-supplemented and unsupplemented animals will show whether the antagonism is reversible and will separate the toxic potential of the agent under study from its effects on nouriture. Such studies will aid in an ultimate assessment of safety in that normal levels of the micronutrient in human populations may be so in excess of absolute requirements as to make this effect of the agent of little consequence. On the other hand, if the nutrient is often at marginal levels in human diets, then clearly its evaluation will have to take this effect into account.

Other specific nutritional problems have been considered in relation to phosphates (23, p. 13) and metals occurring in food (23, p. 14). The problems of the former include alteration of dietary calcium:phosphorus (Ca:P) ratios with consequent complications of, for example, nephrocalcinosis, and the Committee recommended that "further studies should be carried out on the consequences of high dietary intakes of phosphate, with particular reference to the Ca:P ratio." The association between caecal enlargement and nephrocalcinosis also indicates that further complications may arise in relation to calcium and phosphorus absorption due to other food additives.

With respect to the presence of metals in foods, the Committee noted that:
"toxicological evaluation of metals in food calls for carefully balanced consideration of. . . *(inter alia)*

... nutritional requirements, including nutritional interactions with other constituents of food in respect of ... absorption, storage in the body and elimination" and in the case of essential elements, tentative tolerable daily intakes "should not be construed as an indication of any change in recommended daily requirements, but as reflecting permissible human exposure. ..." (23, pp. 14-15).

5.3.3 Health status

Health status of test animals is of key importance in assessing the results of any toxicity study. Health of animals should be routinely monitored during testing. Animals in poor health from a viral or bacterial infection may be especially sensitive to the test substance. Early deaths from infectious diseases may leave insufficient time for chronic toxicity of the test compound such as carcinogenicity to manifest itself. Pathological lesions from infectious disease may also mask compound-related pathology. For example, lung toxicity could be masked by a respiratory infection. The reverse can also occur. Certain respiratory infections in the rat predispose the rat lung to lymphoreticular neoplasms (74: acesulfame potassium, pp. 22-23). Antibiotics and other drugs should not be used unless absolutely necessary to control infections because their use complicates the interpretation of the study.

5.3.4 Study design

When designing toxicity studies on food chemicals where factors of age, nutritional status, and animal health are likely to affect the results, the investigator should design the study appropriately with foreknowledge of these factors, bearing in mind the population likely to be exposed. For example, if the substance is to be used in infant formulas or baby foods, an appropriate animal model to mimic the human infant should be used. The miniature pig may be a useful model in this regard, because it can be bottle-fed and many aspects of its metabolism are similar to those of man.

Most food additives are also consumed by pregnant women, so the factors discussed above with respect to *in utero* exposure should be considered when assessing their safety. Animal studies designed to parallel human exposure should include the important phases of exposure that occur during fetal development and suckling of the infant. Exposure of the test animal can then involve both the parent compound and maternal metabolites that can either cross the placental barrier or enter the mother's milk; it will also permit the assessment of metabolites formed in the developing embryo, which may differ from those formed in the maternal system.

The need for guidelines, as recommended by the twenty-second JECFA (32, p. 30), continues (section 5.3.1.1). The exposure level and its relationship to the no-observed-effect level in animal studies, the type of additive (e.g., nutritive versus non-nutritive), information about whether the additive crosses the placental barrier or is secreted in milk, and other data relating to the reproductive or developmental toxicity of similar compounds should be considered in determining the need for *in utero* studies.

The age of test animals is an important factor to consider when designing carcinogenicity studies (75, pp. 57-107). If the study is terminated too soon, the possibility of detecting carcinogenicity that manifests itself late in the animal's life span is lessened. Conversely, in studies with long exposure times (more than 104 weeks in rats and mice), the background incidence of naturally-occurring tumours may "swamp out" compound-related tumours occurring at some sites. One way to resolve this problem is to add additional groups of animals for interim sacrifice. However, this increases the cost and complexity of the study. Knowledge of the test strain with regard to longevity and tumour incidence is necessary in the design and interpretation of carcinogenicity studies. In any case, final termination should take place while there are still enough survivors among the exposed animals and concomitant controls to make a statistical evaluation.

It generally is not feasible to test all food additives for their effects on all age groups and disease states. In cases where certain populations, such as phenylketonuric patients, are known to be sensitive to a food additive, warning labels or education through other means may be necessary. When substances are included in special medical foods used in treating certain diseases, it would be prudent to examine closely any reported physiological or toxicological effects of the substances, to determine whether they can be safety ingested by the intended population. Specialized testing on animal models may be necessary. If the additive is to be used in infant formulae or "junior foods", this fact should be kept in mind at the time of the safety assessment. This point is often overlooked. Large amounts of such a food additive may be consumed, because the formula may constitute the entire diet of the infant and because infants take in much more food than an adult on a kg body weight basis. If specific subpopulations are identified as being at higher risk than the general population, these groups can be protected by adjusting the ADI to take their special needs into account.

In general, the safety of a food additive, as far as limited special populations not readily identified are concerned, must rely on the conservatisms built into the safety assessment process, the analysis of the data, and the safety factors used in setting an ADI.

5.4 Use of Human Studies in Safety Evaluation

Human studies are not normally included in the data packages that JECFA reviews in its evaluation of new food additives. However, the Committee recognizes the value of human data, has sometimes requested such data, and has always used it in its evaluations when available. Data from controlled human exposure studies are useful in confirming the safety indicated by animal studies after the establishment of ADIs. Such data are also useful in subsequent periodic reviews, and might facilitate a re-evaluation of the safety factors that are applied in calculating ADIs.

Investigation in human subjects was addressed by the WHO Scientific Group on Procedures for Investigating Intentional and Unintentional Food Additives (2, pp. 9-10). The Group felt that "prediction and prevention of possible toxic hazards to the community that might arise from the introduction of a chemical into the environment can be made more certain if information from meaningful studies in human subjects is available." Three particular aspects of toxicology were identified in this connection, "the choice of the most appropriate animal species for. . . the prediction of human responses; secondly, the investigation of a reversible specific effect observed in the most sensitive animal species to determine whether it represents a significant hazard to man; thirdly, the study of effects specific to man."

The Group pointed to:

"the need, at a relatively early stage, to obtain information on the absorption, distribution, metabolism, and elimination of the chemical in human subjects, since this makes it possible to compare this information with that obtained in various animal species and to choose the species that are most likely to have a high predictive value for human responses."

This need has been reiterated by subsequent meetings of JECFA (27, p. 23; 16, p. 31; 32, p. 13) and in WHO Environmental Health Criteria 6 (76). However, the WHO Scientific Group acknowledged that "it is necessary to have adequate short-term toxicological information in several species before even low doses of a new chemical are administered to human subjects" (2, p. 9).

In relation to ascertaining whether the safety margin predicted from animal data is valid, the WHO Scientific Group decided that it might be helpful to administer a chemical to human volunteers, but emphasized the conditions that should be fulfilled with regard to such a study (2, p. 10). *Inter alia*, these conditions include:

(a) The effect or effects studied should be reversible.

(b) The dose levels used should be based on full information of the toxicological properties of the substance in animals.

(c) The investigation should be terminated immediately the effect has been unequivocally demonstrated.

With regard to effects specific to man, the WHO Scientific Group (2, p. 10) considered it unacceptable to study such effects by means of volunteers (in an analogous manner to clinical trials with drugs) but thought that toxicological studies could be made on those who are occupationally exposed to the chemical or in patients suffering from accidental poisoning. A need was identified for "more critical epidemiological and toxicological investigations in such situations." Such studies could be of particular value in relation to hypersensitivity or other idiosyncratic reactions since no suitable animal model has yet been developed. In relation to hypersensitivity, the seventeenth and eighteenth meetings of JECFA (16; 17, p. 10) stated that "no approval would be given for the use of a substance causing serious or widespread hypersensitivity reactions". However, such information can be derived only from studies on human beings.

The WHO Scientific Group has raised an apparent contradiction in its different recommendations with regard to confirming animal studies and investigating effects specific to man. As stated above, the Group recommended that controlled human studies be performed to confirm animal studies, but that it is inappropriate to study effects specific to man by the use of human volunteers. This is all the more perplexing, because controlled human studies, despite their limitations, are the only means available, at present, for studying effects in man that are not observed in animals. JECFA may wish to reconsider the question of using human volunteers to identify specific responses, which would be done only after the usual battery of toxicological investigations had been completed. The words of Paget (77) are cogent in this regard:

"The question is not whether or not human subjects should be used in toxicity experiments but rather whether such chemicals, deemed from animal toxicity studies to be relatively safe, should be released first to controlled, carefully monitored groups of human subjects, instead of being released indiscriminately to large populations with no monitoring and with little or no opportunity to observe adverse effects."

The ethical problems associated with toxicological studies on human beings have been reviewed succinctly in WHO Environmental Health Criteria No. 6 (76, pp. 41-42).

Information relating to human exposure to a food additive during its pre-marketing stage can be obtained through the health monitoring of employees coming into contact with it, either in the laboratory or the manufacturing plant. Because the route of exposure in such a situation is through either contact with the skin or vapour in the lungs, immunological sensitivity and anaphylactoid reactions (mediator-release anaphylaxis-like reactions), often involving histamine release, are the adverse effects most likely to occur. Thus, any observations indicating the potential for these effects should be recorded at the time they are observed.

5.4.1 Epidemiological studies

Most studies of the effects of food additives on human populations are performed after the additive has been placed on the market. In nearly all cases, the impetus for the performance of human studies on a food additive is that the safety of its use has been brought into question for one reason or another. For example, retrospective investigations have revealed effects such as "beer drinkers cardiomyopathy", resulting from exposure to cobalt salts. Adverse findings in these studies may be used for bringing an additive back to JECFA for re-evaluation of its safety.

Epidemiological studies designed to assess the safety of food additives have been performed in several instances, but, generally, definitive results have not been obtained because of the lack of sensitivity of such studies and problems in identifying control populations. For example, long-term low-level nitrite exposure has been very difficult to study epidemiologically because of its ubiquitous nature and the consequent difficulty of finding subpopulations with little or no exposure to nitrite. With saccharin, an extensive data base involving retrospective epidemiological studies and case-control studies has been developed. This data base has been generated using different subpopulations located in different geographical areas, and the results for human bladder cancer have usually been negative (78).

In many cases, the purpose of an epidemiological study is to confirm in human beings a positive finding observed in animals. Thus, when food additives are used extensively and when exposed or unexposed populations cannot be identified, negative results are usually deemed not to be of much value for regulatory purposes. This is because epidemiological studies, are, on the whole, less sensitive than well-designed animal feeding studies. However, when the number of individuals studied becomes very large, this lack of sensitivity is somewhat ameliorated, and safety decisions can be made on the basis of the human studies. For example, in the case of saccharin, negative results in epidemiological studies have been considered important by JECFA for deciding that its continued use is acceptable (1).

Of course, much more can be said about a positive result than a negative one, especially with epidemiological studies, which are usually relatively insensitive. An undetectable adverse effect in a study involving a few thousand individuals could affect a very large number of people in a population of hundreds of millions.

5.4.2 Food intolerance

For the purposes of this discussion, food intolerance is defined as a reproducible, unpleasant reaction to a food or food ingredient, including reactions due to immunological effects, biochemical factors such as enzyme deficiencies, and anaphylactoid reactions, which often include histamine release. Food allergy, sometimes used synonymously with food sensitivity, is a form of food intolerance in which there is evidence of abnormal immunological reaction to the food. Immunological reactions may be further characterized on the basis of the timing of the onset of symptoms following ingestion of the offending food and on the type of response involved. Reactions occurring within minutes to hours of food ingestion are characterized as immediate allergic reactions, which are mediated by Immunoglobulin E (IgE), while reactions beginning several hours to days after food exposure are characterized as delayed allergic, or cell-mediated, reactions.

Various dietary factors may be responsible for food intolerance. These may be naturally-occurring dietary constituents or, in some cases, food additives. Two notable examples of food additives that have been implicated are tartrazine, which may induce urticaria and bronchoconstriction in asthmatic patients, and sodium metabisulfite, which has been associated with bronchospasm, flushing, hypotension, and even death due to anaphalaxis after ingestion by some asthmatic patients. Monosodium glutamate (MSG) gives rise to "Chinese Restaurant Syndrome", manifested largely by violent headache. Certain subpopulations appear to be sensitive to MSG, but the mechanism is unknown. Despite these examples, there is little to suggest that food additives are likely to cause more problems of food intolerance than are components naturally present in food.

Satisfactory animal models to predict food intolerance in human beings have not been developed. At the same time, many difficulties are associated with human studies, and interpretation is difficult at least partly because of the anecdotal nature of much of the evidence. Any interpretation of food intolerance is complicated by psychological factors, making it extremely important that blind trials be performed to assess the nature of the problem.

The most unambiguous method of demonstrating food intolerance is to use challenge feeding in a double-blind study; the diagnosis of food intolerance can only be established if the symptoms disappear with an elimination diet and if a controlled challenge then leads to either recurrence of symptoms or to some other clearly identified change associated with the intolerance. If delayed allergic reactions are being studied, such effects may take several weeks to disappear and then redevelop after challenge feeding. Challenge feeding is most reliable when the ingestion of food is associated with development of symptoms within one to two hours.

No oral food challenge, even if blind, can be perfect for a number of reasons. Presentation of food in capsules may avoid the possibility of reactions in the mouth, pharynx, and oesophagus and may decrease early digestion of the food by salivary enzymes. Small amounts of food may be regurgitated or eructated and identified by taste and smell. Unknown relationships may exist between suspected foods and periods of abstinence from that food before challenge. The presence of other foods eaten with a suspected food may have facilitated or inhibited digestion and absorption (79).

The simplest and most commonly used test for demonstrating IgE antibodies is the direct skin test. However, this test is unreliable as used, because standard dosages of food extracts have not been developed, and, with sufficiently concentrated food extracts, it is possible to evoke positive skin test results in any person tested. Therefore, skin testing results should be verified by testing the extract on non-sensitive individuals (80).

Other methods for testing IgE antibodies to food include the *in vitro* radioimmunoassay and the leukocyte histamine release assay (81). The former assay is limited by an inadequate standardized reporting procedure, making a comparison of results between investigators very difficult (82). The latter has found only limited application, because it requires fresh blood, and only a limited number of allergens can be tested from a single blood aliquot.

Published studies concerning the usefulness of either skin testing or immunoassay to diagnose clinical adverse reactions to food have shown a marked discrepancy in results. In most of these studies, the investigators have relied on the clinical history for determining the false-positive or false-negative rates for skin tests or immunoassays. Such reports are unreliable. Negative results obtained in skin tests or immunoassays should be treated with more confidence than positive results (83, 84).

If evidence of widespread intolerance to a food additive appears in a country that permits its use, procedures should be established for the centralized reporting of such information, if one is not already in place. Medical professionals should be alerted, and appropriate medical tests performed on affected individuals to determine the nature of intolerance. If the problem arises with an additive previously considered and given an ADI by JECFA, ideally the results will be relayed to the Committee so that the safety of the additive can be reconsidered. Remedies may range from no action to a recommendation that the additive be removed from the market. Factors, such as its natural occurrence in food, should be taken into account in such deliberations. Because food intolerance is not spread throughout the general population, but is restricted to small subpopulations or individuals, one of the usual remedies is to label the food containing the additive prominently, so that sensitive individuals can avoid it.

5.5 Setting the ADI

Almost any substance at a high enough test level will produce some adverse effect in animals. Evaluation of safety requires that this potential adverse effect be identified and that adequate toxicological data be available to determine the level at which human exposure to the substance can be considered safe.

At the time of its first meeting, JECFA recognized that the amount of an additive used in food should be established with due attention to "an adequate margin of safety to reduce to a minimum any hazard to health in all groups of consumers" (9, pp. 14-15). The second JECFA, in outlining procedures for the testing of intentional food additives to establish their safety for use, concluded that the results of animal studies can be extrapolated to man, and that

"some margin of safety is desirable to allow for any species difference in susceptibility, the numerical differences between the test animals and the human population exposed to the hazard, the greater variety of complicating disease processes in the human population, the difficulty of estimating the human intake, and the possibility of synergistic action among food additives" (10, p. 17).

This conclusion formed the basis for establishing the "acceptable daily intake", or ADI, which is the end-point of JECFA evaluations for intentional food additives. In the context in which JECFA uses it, the ADI is defined as an estimate (by JECFA) of the amount of a food additive, expressed on a body weight basis, that can be ingested daily over a lifetime without appreciable health risk.

The ADI is expressed in a range, from 0 to an upper limit, which is considered to be the zone of acceptability of the substance. JECFA expresses the ADI in this way to emphasize that the acceptable level it establishes is an upper limit and to encourage the lowest levels of use that are technologically feasible.

Substances that accumulate in the body are not suitable for use as food additives (39, p. 8). Therefore, ADIs are established only for those compounds that are substantially cleared from the body within 24 h. Data packages should include metabolism and excretion studies designed to provide information on the cumulative properties of food additives.

JECFA generally sets the ADI of a food additive on the basis of the highest no-observed-effect level in animal studies. In calculating the ADI, a "safety factor" is applied to the no-observed-effect level to provide a conservative margin of safety on account of the inherent uncertainties in extrapolating animal toxicity data to potential effects in the human being and for variation within the human species. When results from two or more animal studies are available, the ADI is based on the most sensitive animal species, i.e., the species that displayed the toxic effect at the lowest dose, unless metabolic or pharmacokinetic data are available establishing that the test in the other species is more appropriate for man (section 5.5.1).

Generally, the ADI is established on the basis of toxicological information and provides a useful assessment of safety without the need for data on intended or actual use and consumption. However, in setting ADIs, an attempt is made to take account of special subpopulations that may be exposed. Therefore, general information about exposure patterns should be known at the time of the safety assessment (section 5.5.6). For example, if a food additive is to be used in infant formulae, the safety assessment is not complete without looking carefully at safety studies involving exposure to very young animals.

JECFA uses the risk assessment process when setting the ADI, i.e., the level of "no apparent risk" is set on the basis of quantitative extrapolation from animal data to human beings. Generally, JECFA does not undertake risk management, in that it leaves it to national governments to use the quantitative assessments in a manner appropriate to their own situations. However, this has not always been the case, in that sometimes JECFA has taken into consideration, in peripheral ways, benefits (e.g., hydrogen peroxide as an alternative to pasteurization in developing countries (12; 21)) and economic need (e.g, polymer packaging materials that contain potentially hazardous migrants should be limited to situations where no satisfactory alternatives exist (1)). In this context, risk assessment and risk management are more broadly used than, e.g., they are often used in the context of carcinogenesis.

5.5.1 Determination of the no-observed-effect level

A determination of a no-observed-effect level for a study depends primarily on the proper selection of doses, such that the highest dose produces an adverse effect that is not observed at the lowest dose. Several dose levels are used to determine the dose-effect relationship. Knowing the nature of the toxic response to a compound at the high level, a more confident assessment of a no-observed-effect level at the lower test levels can be made by focusing more clearly on the target tissues. Great care must be taken in dose selection, because the no-observed-effect level must be one of the experimental doses; it is not an inherent property of the animal system. For a discussion of items to consider when selecting doses, see reference 75, pp. 9-49.

The following discussion concerns the performance of long-term studies, because these studies are the type most often performed in support of intentional food additives, and they give rise to much controversy. However, 90-day studies are sometimes sufficient for establishing safety, as, for example, with substances that are closely related to food additives of known low toxicity (section 5.5.4). Many of the points discussed below in relation to long-term studies are also appropriate for shorter studies when such studies serve as the basis for safety determinations.

When long-term studies are indicated, short-term range-finding studies should first be performed to ensure the proper selection of dosage regimen. Care must be taken in applying this approach to dose level selection, because doses that produce signs of toxicity in short-term testing may be reversible on more long-term exposure. In such situations, the highest dose selected from range-finding studies may not produce an adverse effect with long-term exposure, precluding a determination of the no-observed-effect level in the longer study (the significance of this transient effect should be taken into account when evaluating the data). A situation of perhaps greater frequency is one in which the dosages in the long-term study are too high, so that, even the lowest dose results in adverse effects, and a no-observed-effect level cannot be established.

Ideally, in a long-term study, the high-dose level should be sufficiently high to elicit signs of toxicity without causing excessive mortality or some exaggerated pharmacological effect, such as sedation. Although doses of non-nutritive additives as high as 5% of the total diet do not always produce adverse effects, higher doses should not be tested, because they may produce a significant nutritional imbalance. Therefore, if no adverse effects are observed at 5% of the diet, this dose should be considered the no-observed-effect level. On the other hand, nutritive additives may be fed at higher doses as long as the nutritional balance is effectively preserved in both the test animal and controls (section 6.2.3).

Ordinarily, the middle dose should be selected to be sufficiently high to elicit minimal toxic effects or it should be set midway between the high and low doses. However, if significant differences exist in the pharmacokinetic or metabolic profile of the test substance between the high and low doses, then an additional dose should be included in the study to provide more assessment points.

The lowest dose should not interfere with morphology, development, normal growth, or longevity or produce adverse functional alterations.

The determination of an adverse effect in a particular study depends on the doses tested, the types of parameters measured, and the ability to distinguish between real adverse effects and false positives. If, for example, only a slight change in a particular parameter is noted at the highest dose that is not observed at the lower doses, then it is difficult to distinguish between a real adverse effect and a spurious positive finding. In addition, a reduction in body-weight gain coupled with decreased food consumption is difficult to interpret as an adverse effect, because palatability of the chow might be affected by the presence of high levels of the test compound. However, as noted in section 5.1.1, generalized decrement in weight gain has sometimes been used for setting an effect level in the absence of other toxic manifestations.

When two or more studies are performed on an additive in different animal species, no-observed-effect levels are calculated from each study. The overall no-observed-effect level used for calculating the ADI is the no-observed-effect level from the animal study that displayed a toxic effect at the lowest dose. The species on which this study was performed is then considered to be the most sensitive species. This approach is reasonable when the animal studies are of similar length (in relation to the expected life span of the species) and quality, and no other data relating to this issue are available. However, if the quality of one study is obviously superior to the others and/or the studies differ with respect to length (long-term versus short-term), extra weight should be given to the longer better-quality studies when determining the overall no-observed-effect level. If metabolic and pharmacokinetic data are available, the species most similar to man with respect to the toxic effect should be used in calculating the overall no-observed-effect level, rather than the most sensitive species.

5.5.2   Use of the safety factor

The safety factor has been used by JECFA, since its inception. It is intended to provide an adequate margin of safety for the consumer by assuming that the human being is 10 times more sensitive than the test animal and that the difference of sensitivity within the human population is in a 10-fold range. In determining an ADI, a safety factor is applied to the no-observed-effect level determined in an appropriate animal study.

JECFA traditionally uses a safety factor of 100 (10 x 10) in setting ADIs based on long-term animal studies, i.e., the no-observed-effect level is divided by 100 to calculate the ADI for an additive. The no-observed-effect level is usually expressed in terms of mg compound per kg body weight per day, and the ADI is expressed in the same units. A food additive is considered safe for its intended use if its human exposure is less than, or is approximately, the same as the ADI. The ADI generally includes both its natural occurrence and deliberate addition to food (17, pp. 8-10), except when the substance occurs naturally in a chemical form different from that employed as a food additive, or when its natural occurrence was not considered when setting the ADI and the substance naturally present in the diet contributes significantly to its total intake (as with nitrates). Because in most cases, data are extrapolated from life-time animal studies, the ADI relates to life-time use and provides a margin of safety large enough for toxicologists not to be particularly concerned about short-term use at exposure levels exceeding the ADI, providing the average intake over longer periods of time does not exceed it.

National governments are responsible for regulating food additives in such a way that consumption from natural occurrence and deliberate addition to food does not exceed the ADI for each additive that is permitted. As stated by the WHO Scientific Group on Procedures for Investigating Intentional and Unintentional Food Additives (2, p. 6), "it is desirable that national governments should maintain a check on the total intake of each food additive, based on national dietary surveys, to determine whether the total load in the diet approaches the acceptable daily intake." Individual governments have the discretion of determining whether they will base their regulatory decisions on the "average" consumer or the "high" consumer of food additives.

A safety factor of 100 should not be considered immutable. When setting the ADI, various test data and judgemental factors should be considered. These include:

(a) Inadequate data base

In this case, a larger safety factor may be appropriate (section 5.5.5).

(b) <u>Reversibility of the observed effect in embryotoxicity studies</u>

If irreversible developmental effects, such as skeletal abnormalities (as opposed to retarded skeletal growth), are seen in the fetuses of animals administered the substance in utero, a study on a second species is indicated. If similar irreversible effects are not confirmed in the second animal species, pharmacokinetic studies would be useful to determine relevance to human beings. Judgement would then be needed to set an appropriate safety factor. If frank teratogenic effects are observed in both studies, judgement would be needed to decide whether either a larger safety factor should be considered or it should be recognized that the use of the substance as a food additive is not appropriate. If only reversible developmental effects are seen, such as retarded skeletal and soft tissue development or decreased fetal weight, the usual safety factor of 100 may be applied.

(c) <u>Age-related effects in reproduction studies</u>

Such studies may demonstrate different toxic responses in young animals compared with older ones. Metabolic studies may demonstrate that the differences in sensitivity are due to such factors as incomplete development of enzyme systems used for metabolizing xenobiotic compounds or differences in intestinal flora. Safety factors should be set on the basis of the target population. If young children are likely to consume the additive, the ADI should be based on the no-observed-effect level from the phase of the study in which young animals were exposed, if the no-observed-effect level was lower than in the adult phase. If, on the other hand, it is shown that children will not be exposed to the additive, it may be appropriate to set the ADI on the basis of the no-observed-effect level established in the adult phase of the study.

(d) <u>Finding of carcinogenicity</u>

Carcinogens vary in the magnitude of risk they present for man, because they act via different mechanisms. Even though no basis exists for the exact extrapolation of risk from experimental animals to man, the degrees of risk from different carcinogens can often be inferred from the data. However, with the present state of knowledge, it would be appropriate to consider the use of a carcinogenic substance as an intentional food additive only under very restricted circumstances. For example, if cancer is shown to be a secondary effect, such as bladder tumours occurring secondary to the induction of bladder stones, and there is evidence of a threshold below which the additive is safe, then it would be appropriate to use a safety factor for determining the safe level of use of the additive. Under extenuating circumstances, such as an unambiguous demonstration that the health benefits exceed the risk, it may also be possible to use a carcinogenic additive.

(e) If reasons exist for setting a lower safety factor

If toxicity and dose-response effects in human beings are known, such data should take precedence over extrapolation from animal studies; a 10-fold safety factor would be appropriate if there is no evidence that human sensitivity to the agent varies more than 10-fold among individuals. A lower safety factor may also be appropriate when the additive is similar to traditional foods, is metabolized into normal body constituents, and/or lacks overt toxicity. Also, a 100-fold safety factor often would not provide a high enough level of nutrients required to satisfy nutritional needs and to maintain health (toxicity for some essential nutrients such as Vitamin A, Vitamin D, certain essential amino acids, and iron may be reached at levels less than 10 times higher than those recommended for optimal nutrition). A substance that serves as a significant source of energy in the human diet obviously cannot fit into the constraints of a 100-fold safety factor.

The use of standardized safety factors based on no-observed-effect levels for establishing the acceptable level of use of food additives is a crude procedure, given the known wide variability in toxic responses. For example, the nature of the dose response usually is not used. In part, this is a reflection of the fact that good dose-response data are not available for many compounds. Attempts to use the dose-response behaviour of compounds in establishing quantitative end-points must contend with this limitation.

In the broadest sense, the procedures used by JECFA take into account the nature of the biological effects observed in animal bioassays only to the extent that a distinction is made between carcinogens and non-carcinogens, i.e., ADIs are established for non-carcinogens, while most carcinogens are considered to unacceptable for use as intentional food additives. Otherwise, the nature of the observed effect is not an explicit component of the quantitative assessment of food additives. However, the nature of the effect and a determination of its significance are often implicitly considered by scientists, when reviewing the data.

JECFA should take these and other factors into account when determining acceptable daily intakes of food additives. However, in situations in which little is known beyond the empirical finding of toxicity in animal studies, the traditional approach for calculating ADIs would seem to be appropriate. This may be an issue for future consideration by JECFA.

5.5.3 Toxicological versus physiological responses

When analysing a toxicological study and setting a no-observed-effect level, a distinction must be drawn between reversible changes that are due entirely to normal physiological processes or homeostasis-maintaining mechanisms, and to toxic responses themselves (section 5.1). Examples of the former include: laxative effects from osmotic or faecal overload, liver hypertrophy and microsomal enzyme induction from high doses of substances metabolized by the liver, decreased body weight gain or caecal enlargement from high levels of non-nutritive substances, alteration in renal weight that is directly related to the amount of water being processed by the kidney, and decreased growth rate and food consumption related to the dietary administration of an unpalatable substance. However, care must be taken in interpreting these changes, and they should not automatically be dismissed as being unimportant from a toxicological point of view. For example, microsomal enzyme induction in the liver may result in alterations in the metabolism of compounds unrelated to the administered substance, which could result in a toxic effect. A decrease in the rate of body-weight gain coupled with a corresponding reduction of food intake could be due to toxic anorexia, rather than a palatability defect.

The dose at which the effect occurs should be compared with the amount of the substance consumed by human beings. Thus, it would ordinarily be acceptable to permit the use of a substance that causes diarrhoea only at very high levels of consumption in rats, but the use of such a substance should be severely restricted or not permitted if it causes diarrhoea at normal levels of consumption in human beings. Sometimes, physiological adaptation may progress through overload to frank toxicity.

Further studies are indicated in situations in which it is difficult to draw a clear distinction between a toxic and a physiological response. Special studies such as paired feeding, caloric balance comparisons between food consumption and body-weight gain, or, in the case of reproduction studies, cross fostering, can be performed to decide issues such as reduced food intake and reduced body-weight gain related to unpalatable test substances. Metabolic and pharmacokinetic studies may be of use in providing information on the distribution of the test compound and its metabolites or the dose at which a change in metabolism occurs.

5.5.4 Group ADIs

If several compounds that display similar toxic effects are to be considered for use as food additives, it may be appropriate in establishing an ADI to consider the group of compounds in order to limit their cumulative intake. For this procedure to be feasible, the additives must be in the same range of toxic potency. Flexibility should be used in determining which no-observed-effect level is to be used in calculating the ADI. In some cases, the average no-observed-effect level for all the compounds in the group may be used for calculating the group ADI. A more conservative approach is to base the group ADI on the compound with the lowest no-observed-effect level. The relative quality and length of studies on the various compounds should be considered when setting the group ADI. When the no-observed-effect level for one of the compounds is out of line with the others in the group, it should be treated separately.

When considering the use of a substance that is a member of a series of compounds that are very closely related chemically (e.g., fatty acids), but for which toxicological information is limited, it may be possible to base its evaluation on the group ADI established for the series of compounds. This procedure can only be followed if a great deal of toxicological information is available on at least one member of the series and if the known toxic properties of the various compounds fall along a well-defined continuum. Interpolation, but not extrapolation, can be performed by this procedure. The use of this procedure by JECFA represents one of the few situations in which the Committee has used structure/activity relationships in its safety assessments.

In some instances, group ADIs can be established primarily on the basis of metabolic information. For example, the safety of esters used as food flavours could be assessed on the basis of toxicological information on their constituent acids and alcohols, provided that it is shown that they are quantitatively hydrolysed in the gut.

The calculation of a group ADI is also appropriate for compounds that cause additive physiological or toxic effects, even if they are not closely related chemically. For example, it may be appropriate to establish a group ADI for additives such as bulk sweeteners that are poorly absorbed and cause laxation.

5.5.5 Special situations

There are occasions when JECFA considers the use of an ADI in numerical terms not to be appropriate. This situation arises when the estimated consumption of the additive is expected to be well below any numerical value that would ordinarily be assigned to it. Under such circumstances, JECFA uses the term "ADI not specified". The Committee defines this term to mean that, on the basis of available data (chemical, biochemical, toxicological, and other), the total daily intake of the substance, arising from its use at the levels necessary to achieve the desired effect and from its acceptable background in food, does not, in the opinion of the Committee, represent a hazard to health. For that reason, and for the reasons stated in the individual evaluations, the establishment of an ADI in numerical form is not deemed necessary (e.g., 1, Annex II). An additive meeting this criterion must be used within the bounds of good manufacturing practice, i.e., it should be technologically efficacious and should be used at the lowest level necessary to achieve this effect, it should not conceal inferior food quality or adulteration, and it should not create a nutritional imbalance (16, pp. 10-11). That the background occurrence of the chemical must be taken into account in the evaluation of its safety was articulated by the WHO Scientific Group on Procedures for Investigating Intentional and Unintentional Food Additives (2, p. 7).

JECFA has encountered several situations in which either the body of data before it on a new additive was limited, or the safety of a food additive for which the Committee previously assigned an ADI was brought into question by the generation of new data. When the Committee feels confident that the use of the substance is safe over the relatively short period of time required to generate and evaluate further safety data, but is not confident that its use is safe over a lifetime, it often establishes a "temporary" ADI, pending the submission of appropriate data to resolve the safety issue on a timetable established by JECFA. When establishing a temporary ADI, the Committee often uses a higher-than-usual safety factor, usually increasing it by a factor of 2. The additional biochemical and toxicological data required for the establishment of an ADI are clearly stated, and a review of these new data is conducted before the expiration of the provisional period.

This approach seems to have worked reasonably well in practice in that it has encouraged necessary research without creating any known safety problems. In many cases, long-term studies are requested, but timetables are not met, which means that JECFA has had to extend temporary ADIs for long periods of time. JECFA has withdrawn ADIs, in instances where data were not forthcoming, as a safety precaution.

5.5.6 *Comparing the ADI with potential exposure*

When establishing ADIs, collateral exposure information is often useful for determining the relationship between the two values. The agreement between exposure and acceptable daily intake helps to determine whether an "ADI not specified" should be established. Exposure information is also indispensable when:

(a) performing risk assessments for food contaminants and processing aids; and

(b) assessing the safety of added substances that may be naturally present in food to determine their relative contributions to the diet (17, pp. 8-10).

In order to accurately compare exposure and acceptable intake, similar assumptions should be used for making each estimate or, at least, the differences and similarities in the estimates should be understood. For example, if an ADI is computed from lifetime dosage, then the estimated human exposure should represent lifetime exposure to the additive. Sometimes, acceptable intakes are computed for specific age groups or for certain dosage conditions when short-term exposure should be limited, such as with certain food additives that cause laxative effects at high dose levels. Under such circumstances, the estimated human exposure should represent the same age group or dosage conditions. In practice, however, estimates of exposure do not represent exposure for individual consumers in the same way that toxicological data represent dosages for individual animals. Data bases on food and food additive intakes provide composite data for subpopulations, such as the average dietary habits of a particular nation's population.

For effective comparison of exposure estimates with acceptable intakes, the assumptions used to compute exposure estimates should always be stated. Data on the functional use(s) of intentional food additives and information on approaches used to compute intake estimates, such as analytical studies on food constituents or migration (carry-over) models for certain contaminant situations, should be provided if possible.

Each estimate of exposure represents a facet of actual human exposure and, thus, each estimate represents useful scientific data. However, it is not possible to describe specific procedures for estimating exposure for all food additive and contaminant situations. However, JECFA is able to provide guidance by describing the types of estimation procedures that have been accepted by previous Committees, which are discussed in section 3.1.1

6. PRINCIPLES RELATED TO SPECIFIC GROUPS OF SUBSTANCES

6.1 Substances Consumed in Small Amounts

Many of the substances that come before JECFA for its evaluation are present in food in only trace amounts. Testing requirements generally take these low exposures into account (section 3.1). However, as discussed below, special safety concerns are raised by the use of many of these substances, despite the low exposures to both the parent compounds and their residues.

In some cases, these substances have no technological function in the food itself. Some are used in food processing. For example, residues from extraction solvents used *inter alia* in extracting fats and oils, in defatting fish and other meals, and in decaffeinating coffee and tea may be present in the final food product because of incomplete removal. The same is true for enzymes and immobilizing agents (and their residues) used in immobilized enzyme preparations. Residues arising from the use of xenobiotic anabolic agents and from the use of packaging materials may also occur in food.

Residues belonging to all these classes of substances have been evaluated by JECFA, and the Committee has developed guidelines concerning their safety evaluation. The guidelines, which are reproduced in Annex III, are intended to serve as examples of guidance by JECFA for evaluating these specific categories of substances. Further discussion of such substances and others follows in section 6.1.1.

Flavouring agents constitute a category of substances that have a functional effect in food, but are generally added in small amounts. The safety evaluation of flavours has presented special problems for JECFA, and these are discussed in detail in section 6.1.2.

6.1.1 Food contaminants

JECFA has considered the presence of food contaminants on many occasions since 1972, when mercury, lead, and cadmium were first assessed (60, pp. 11-24). These food contaminants have included, in addition to heavy metals, environmental contaminants such as mycotoxins, impurities arising in food additives, solvents used in food processing, packaging material migrants, and residues arising from the use of animal feed additives and/or veterinary drugs. Each of these classes of food contaminants possesses its own unique characteristics and evaluation requirements. Thus, JECFA has recognized through the years that evaluation principles should pertain to classes or groups of contaminants rather than to food contaminants *in toto*. JECFA has published guidelines, which are reproduced in Annex III, for the evaluation of various classes of contaminants; these guidelines are still valid.

At the time that JECFA considered mercury, cadmium, and lead, in 1972, it established the concept of "provisional tolerable weekly intake" (PTWI), which is a departure from the traditional ADI concept (60, pp. 9-11). JECFA has continued to use this concept, with some modifications, ever since.

ADIs are intended to be used in allocating the acceptable amounts of an additive for necessary technological purposes. Obviously, trace contaminants have no intended function, so the term "tolerable" was seen as a more appropriate term than "acceptable", which signifies permissibility rather than acceptability for the intake of contaminants unavoidably associated with the consumption of otherwise wholesome and nutritious foods.

In this convention, tolerable intakes are expressed on a weekly basis, because the contaminants given this designation may accumulate within the body over a period of time. On any particular day, consumption of food containing above-average levels of the contaminant may exceed the proportionate share of its weekly tolerable intake. JECFA's assessment takes into account such daily variations, its real concern being prolonged exposure to the contaminant, because of its ability to accumulate within the body over a period of time.

The use of the term "provisional" expresses the tentative nature of the evaluation, in view of the paucity of reliable data on the consequences of human exposure at levels approaching those with which JECFA is concerned.

A tolerable intake, as defined above, represents the maximum acceptable level of a contaminant in the diet; the goal should be to limit exposure to the maximum feasible extent, consistent with the PTWI. However, potent carcinogens, such as certain mycotoxins, cannot be made to fit within the confines of a PTWI because, using the traditional approach, safe levels cannot be set. JECFA addressed this issue in 1978 and introduced the concept of an "irreducible level", which it defined as "that concentration of a substance which cannot be eliminated from a food without involving the discarding of that food altogether, severely compromising the ultimate availability of major food supplies" (32, pp. 14-15).

Another JECFA end-point, the "provisional maximum tolerable daily intake" (PMTDI) has been established for food contaminants that are not known to accumulate in the body, such as tin (23), arsenic (24), and styrene (1). The value assigned to the PMTDI represents permissible human exposure as a result of the natural occurrence of the substance in food and in drinking-water.

In 1982, JECFA decided to change the methodology of assessment for trace elements that are both essential nutrients and unavoidable constituents of food, such as copper and zinc. The Committee concluded that, in such situations, the use of one number for expression of the tolerable intake was not sufficiently informative, so expression in a range was instituted (23, Annex II). The lower value represents the level of essentiality and the upper value the PMTDI. Thus, the upper value should not be construed as representing its normal daily requirement.

With the use of increasingly sensitive analytical techniques, it is becoming clear that many food additives also contain trace levels of carcinogenic contaminants. JECFA addressed this issue in 1984 when it considered the low-level migration of carcinogenic contaminants from food packaging materials (1). The Committee did not consider it appropriate to allocate ADIs on the basis of the information available. For further evaluation, the twenty-eighth JECFA stated that it would need the following information:

(a) the lowest levels of potential migrants from within the polymeric system(s) that are technologically attainable with improved manufacturing processes for food-contact materials;

(b) the resulting levels of the migrants in foods;

(c) the intake of the foods; and

(d) the most appropriate statistical design that will enable the implications for health to be interpreted from adequate and relevant toxicological data.

In the meantime, the Committee recommended that "human exposure to migrants from food-contact materials be restricted to the lowest levels technologically attainable" (1, p. 23).

6.1.2   Food flavouring agents

The special problems associated with the safety evaluation of food flavouring agents have been apparent to JECFA for a considerable time and the tenth meeting of the Committee recommended that "further meetings of JECFA should be convened to draw up specifications for flavouring substances, . . . used as food additives, and to evaluate the toxicological hazards involved in their use" (29). The Committee referred to the extent and magnitude of these problems in the report of the eleventh meeting (41) but, despite the time that has elapsed, the problems remain largely unresolved. Much of the difficulty arises from the fact that a very large number of substances are used as food flavouring agents, many of which occur in natural products, but their level of use is generally low and self-limiting. The compounds and other materials (extracts, oleoresins, essential oils) used to impart flavour to foods can be classified into four groups (19), viz:

(a) artificial substances unlikely to occur naturally in food;

(b) natural materials not normally consumed as food, their derived products, and the equivalent nature-identical flavourings;

(c) herbs and spices, their derived products, and the equivalent nature-identical flavourings; and

(d) natural flavouring substances obtained from vegetable and animal products and normally consumed as food whether processed or not, and their synthetic equivalents.

Most of these food flavouring materials have not been subjected to detailed and comprehensive toxicity tests, though with the flavouring agents in classes (c) and (d) above there may be a long history of use and limited evidence of safety-in-use. In some cases, there may also be evidence of adverse consequences of human exposure such as hypersensitivity and idiosyncratic intolerance, which have been observed with capsaicin, zingiberin, and menthol. However, the natural origin of a food flavouring agent is no guarantee of safety, nor does traditional use of a material constitute unequivocal evidence of safety; the flavour saffrole is of natural origin (oil of sassafras) and had a long history of use before it was demonstrated to be hepatoxic and carcinogenic (22). Consequently, natural flavouring compounds cannot be exempted from the toxicological evaluation applicable to food flavouring agents in general. Conversely, there is no basis for assuming that compounds that interact with gustatory or olfactory receptors are more likely to interact with other physiological receptor sites than non-flavoured compounds. In principle, therefore, the safety evaluation of food flavouring compounds is similar to that for other food additives, and this fact was recognized in the twentieth report of JECFA (19). However, this Committee concluded that evaluation should be flexible and "may require extensive toxicological testing or be made simply from available data". In the evaluation process, the work of other bodies, such as national governments and the Council of Europe, should be considered.

In view of the very large number of substances used as food flavouring agents and the fact that they are generally applied in low and self-timing concentrations in foods, it is considered impractical and unreasonable to require that each food flavouring material be subjected to the same and extensive toxicological evaluation within a reasonable period (32). Furthermore, human data may be available that have a bearing on the extent of toxicity testing required and the urgency with which data from such tests are needed. Thus, there is a need to develop a rational basis for allocating priorities for the testing and safety evaluation of food flavouring agents and for determining the extent of the testing required. Previous Committees have drawn attention to this need on several occasions (19, 32, 31, 24). Several of the factors that influence the allocation of priorities are also relevant to the extent of testing required and some of these are discussed in general terms elsewhere in this report (sections 3, 5.2, and 5.4). The following discussion is specifically concerned with food flavouring compounds.

Factors that should be considered in the allocation of priorities and the determination of the extent of testing required include:

(a) nature and source of the material;

(b) data on usage and on the extent and frequency of exposure of the average consumer and of subpopulations, who may be highly exposed (including exposure from natural sources such as herbs and spices but excluding bizarre eating behaviour);

(c) structure/activity relationships and the similarity of compounds of known toxicological and biochemical properties;

(d) similarity to compounds of known biological activity in metabolism and pharmacokinetics;

(e) information from short-term tests for mutagenicity and clastogenicity;

(f) prior experience of human use; and

(g) toxicological status/regulatory status previously determined by national regulatory agencies or supranational organizations such as the European Economic Community (EEC) Scientific Committee for Foods and the International Agency for Research on Cancer (IARC).

Information on the nature and source of the flavouring material is clearly necessary in order to assess overall exposure to its components, and such information may assist in the allocation of priorities and level of testing required. Data obtained on synthetic compounds may assist in the evaluation of limited data on extracts or essential oils that contain these compounds naturally. For example, knowledge of the composition of an extract or essential oil may indicate that no further testing is necessary; conversely, the presence of a known toxic compound as a major component in an extract or essential oil (e.g., saffrole in oil of sassafras, thujone in oil of wormwood, tansy, etc.) may alter the priority given to that extract or essential oil and determine the nature of specific toxicity tests required. It must be stressed that any assessment of priorities will depend on the availability of adequate specifications which, in the case of complex mixtures such as spice extracts, may include maximum limits for known toxic constituents.

In considering the extent of exposure, the eleventh meeting of JECFA (41) suggested that flavouring compounds with an estimated *per capita* consumption exceeding 3.65 mg per annum (sustained intake exceeding 10 µg per day), and/or use in food at a level higher than 10 mg/kg food, should be given priority. However, exposure must be considered in the light of other available information and it is inappropriate to classify flavouring compounds for priority evaluation solely on the basis of exposure levels.

In the absence of other data or considerations, a combination of exposure levels and structure/activity relationships can be used to establish both priorities and requirements for the level of testing. A number of schemes that take these factors into account have been proposed to accomplish this (34 - 38).

The general principles regarding structure/activity relationships are discussed in section 3.1.2. With particular reference to food flavouring compounds, these principles have been applied already by JECFA in a limited number of cases, both in relation to high levels of concern (flavouring compounds structurally-related to saffrole) and acceptance of limited toxicological data (simple structural analogues, homologues, or derivatives such as esters). In the latter case, detailed toxicological data on one member of a group of related compounds together with metabolic and pharmacokinetic data can be used to allocate an ADI or group ADI to structurally related compounds without the need for further testing. On this basis, a substantial number of esters used as food flavouring agents would warrant the allocation of a low priority for testing and acceptance for an ADI on the basis of metabolic studies alone. This would be true in cases in which flavouring compounds are shown to be rapidly and quantitatively hydrolysed to toxicologically known alcohols and acids, the safety of which have been established.

The use of short-term tests for mutagenicity/clastogenicity cannot at present be considered a substitute for carcinogenicity bioassays and hence negative results in such tests need to be considered in the light of the total toxicological data base in allocating priorities. In general, such negative results would not justify waiving the requirement for long-term carcinogenicity studies on food flavouring compounds, but they may be useful supplementary information to that discussed above. Conversely, positive results in a suitable battery of short-term mutagenicity/clastogenicity tests would indicate a higher priority for more detailed testing.

Prior experience of human use will influence the allocation of priorities, as indicated earlier, and even the limited evidence of safety-in-use may support judgements based on the other criteria discussed above, to allocate a low priority to testing or to accept limited toxicological data. Conversely, observations of human idiosyncratic intolerance and/or allergies would indicate a need for adequate information relating to the extent and severity of the problem.

Systematic treatment of the above data (structure/activity relationships, exposure, and human data) and the application of reasonable criteria for the adequacy of existing data, could provide a useful approach in focusing effort on substances that should receive priority from a scientific point of view. Other criteria that influence the priority that JECFA allocates to food flavours include requests from, or previous evaluation by, national governments, the European Economic Community Scientific Committee for Foods, the Codex Alimentarius Committee on Food Additives, and the need to re-evaluate previously established temporary ADIs. Earlier JECFAs have recommended the setting up of a working group of experts specifically to consider the allocation of priorities for the testing and evaluation of food flavours, and this recommendation still appears to be appropriate. There is an implicit need for adequate resources to perform this task and cooperation with other interested groups to avoid wasteful duplication of effort.

6.2 Substances Consumed in Large Amounts

The safety assessment of substances that are consumed in relatively large amounts presents a number of special problems. Such materials include defined chemical substances such as sorbitol and xylitol (23, 24), modified food ingredients such as modified starches (23), and foods from novel sources.

The safety assessment of such substances should differ from that of other food additives, such as colouring and flavouring agents, and antioxidants, for the following reasons:

(a) Many will have a high daily intake and, thus, minor constituents and processing impurities assume greater-than-usual significance.

(b) Even though they are often structurally similar or even identical to natural products used as food and thus may appear to be of low toxicity, many may require extensive toxicity testing, because of their high daily intake.

(c) Some may be metabolized into normal body constituents.

(d) Some substances, particularly foods from novel sources, may replace traditional foods of nutritional importance in the diet.

(e) Many are complex mixtures rather than defined chemical substances.

(f) The difference between the quantity that can be fed to animals in feeding tests and the amount consumed by human beings is often relatively small.

6.2.1 Chemical composition, specifications, and impurities

Thorough chemical analysis should be performed on high-consumption substances to measure potential impurities and to provide information on nutritional adequacy, especially when such substances replace traditional food.

It is not possible to provide a checklist of necessary chemical studies to cover all high-consumption compounds. However, the substance should be subjected to a full proximate analysis and particular attention should be paid to the points discussed in the following paragraphs.

Because the intake of undesirable impurities concomitant with the intake of high-consumption materials is potentially high, special effort should be made to identify the impurities. Information on the production process, including the materials and procedures involved, will point to the types of contaminants for which limits may need to be specified. The specifications should be accompanied by details of product variability and of the analytical methods used to check the specifications and details of the sampling protocols. If the substance is so complex that comprehensive product specifications on chemical composition are impracticable (as it might be for a microbial protein), the description of the substance in the specifications may include relevant aspects of its manufacturing process. If manufacturing data are based on production on a pilot scale, the manufacturer should demonstrate that, when produced in a large-scale plant, the substance will meet the specifications established on the basis of pilot data.

The permissible limits for impurities may in some cases correspond to the levels accepted for natural foods that have similar structure or function, or that are intended to be replaced by the new material. If the substance is prepared by a biological process, special attention should be paid to the possible occurrence of natural toxins (e.g., mycotoxins).

The substance should be analysed for the presence of toxic metals. Depending on the intended use, analysis for metals of nutritional significance may also be appropriate.

If the nature of the substance or manufacturing process indicates the possible presence of naturally-occurring or adventitious antinutritional factors (phytate, trypsin inhibitors, etc.), or toxins (haemagglutinins, mycotoxins, nicotine, etc.), the product should be analysed for them specifically. Biological tests, either as part of the nutritional evaluation in the case of enzyme inhibitors or more specifically as part of a mycotoxin screening programme, will provide useful back-up evidence concerning the presence or absence of these contaminants.

Finally, if under the intended conditions of use the substance may be unstable or is likely to interact chemically with other food components (e.g., degradation or rearrangement of the substance during heat processing), data should be provided on its stability and reactivity. The various tests should be conducted under conditions relevant to the use of the substance (e.g., at the acidity and temperature of the environment and in the presence of other compounds that may react).

6.2.2 Nutritional studies

With some substances, particularly with novel foods, nutritional studies may be necessary to forecast the likely impact of their introduction on the nutritional status of consumers. In addition to affecting the nutritional content of the diet, such substances may influence the biological availability of nutrients in the diet. The nutritional consequences of the introduction of such a substance in the diet can only be judged in the light of information about its intended use. As much information as possible should therefore be obtained about potential markets and uses, and the likely maximum consumption by particular subpopulations should be estimated.

6.2.3 Toxicity studies

When testing high-consumption additives, animals should generally be fed the highest levels possible consistent with palatability and nutritional status. Therefore, before beginning such studies, it is desirable to investigate the palatability of the test diet in the test animals. If a palatability problem is encountered, it may be necessary to increase the amount of the test substance to the required level gradually. Paired-feeding techniques should be used, if the problem cannot be overcome. It should always be borne in mind that there are practical limits to the amounts of certain foods that can be added to animal diets without adversely affecting the animal's nutrition and health.

To ensure that the nutritional status of the test animal is not distorted, the test and control diets should have the same nutritive value in terms of both macronutrients (e.g., protein, fat, carbohydrate, and total calories) and micronutrients (e.g., vitamins and minerals). When feeding substances at high levels, it is usually advisable to formulate diets from individual ingredients (rather than adding the test material to a standard laboratory diet) to provide the same nutrient levels in the control and test diets. Comprehensive nutrient analyses of the test and control diets should be performed to ensure that they are comparable nutritionally. Sometimes nutritional studies are advisable before toxicological studies are performed to ensure that test diets are correctly balanced. Without due regard to nutritional balance, excessive exposure may investigate the toxicological end-points of long-term dietary imbalance rather than the toxic effects of the substance.

The establishment of a precise no-observed-effect level will not usually be feasible on account of the relative non-toxicity of high-consumption additives and the impracticability of achieving an adequate safety margin between the no-observed-effect level in animals and the expected consumption of such substances by human beings. Therefore, it is particularly important that the variables for assessing the safety of the substance, such as body weight, food and water consumption, haematological parameters, ophthalmology, blood chemistry, urine analysis, faecal analysis, mineral and vitamin excretory levels, etc., are chosen carefully to include monitoring of all its possible toxic effects.

Metabolic studies are useful and necessary for assessing the safety of high-consumption additives. With complex mixtures, studies on the metabolic fate of every constituent would be impracticable. However, if contaminants or minor components are suspected as the cause of toxicity, their metabolism should be investigated. Consideration should also be given to the effects that the new constituents may have on the ability of the host to withstand other toxic agents, for example, effects on the metabolism of xenobiotic compounds. If the material, or a major component of it, consists of a new chemical compound that does not normally occur in the diet (e.g., a novel carbohydrate), studies of the metabolic fate of the new compound would be appropriate.

If biochemical and metabolic studies show that the test material is completely broken down in the food or in the gastrointestinal tract to substances that are common dietary or body constituents, then other toxicity studies may not be necessary. The results of metabolic studies can stand on their own if it is shown: that breakdown into these common constituents occurs under the conditions of normal consumption of the material, that the material contributes only a small proportion of these common constituents in the daily diet, and that side reactions giving rise to toxic products do not occur.

Analysis of urine and faeces may provide important information relating to changes in normal excretory functions caused by the test substance. For example, the gut flora may be altered, or preferential loss of a mineral or vitamin may occur, resulting in detrimental effects on the health of the test animals. If the substance is incompletely or not degraded by the digestive enzymes of the stomach or the small intestine, appreciable concentrations may be found in the faeces or in the distal gut compartments. As a result, changes in the absorption of dietary constituents or changes in the composition and metabolic activity of the intestinal flora may be observed. Because of species-dependent anatomical differences in the digestive tract and because of considerable differences in the composition of the basal diet, such effects may occur only in man but not in rodents, or *vice versa*. Therefore, short-term biochemical studies should be performed in animals and man (if possible) in which variables likely to be affected by the test compound are examined in detail. It is especially important to investigate questions relating to whether the eventual effects are progressive or transient and whether they occur in subjects exposed to the compound for the first time and/or in subjects adapted to a daily intake of the substance. Clearly, no standard design for such studies can be devised. Only a thorough knowledge of the nutritional and biochemical literature can serve as a guideline.

When establishing an ADI, the traditional concept of a 100-fold safety factor cannot operate when the human consumption level is high and feeding studies do not produce adverse effects (except for effects arising from the physical properties of the additive, such as its bulk and hydrophilicity), even when the substance is added to the diet in the maximum possible proportion, consistent with reasonable nutrition. In such cases, new approaches are indicated, including setting the ADI on the basis of a smaller safety factor, which may be permissible when factors such as similarity to traditional foods, metabolism into normal body constituents, lack of overt toxicity, etc., are considered. For a compound, such as a bulking agent, that may influence the nutritional balance or the digestive physiology by its mere bulk and which may be absorbed from the gut only incompletely or not at all, it may be more appropriate to express the dosage level in terms of the percentage inclusion in the diet. In using this approach, a direct comparison between the proportion in the human diet, with a small safety factor, can be made. If several similar types of compounds are likely to be consumed, a group ADI (limiting the cumulative intake) should be allocated.

The results of human studies, which are discussed in relation to novel foods in section 6.2.4, may permit the use of a lower safety factor than that obtained from animal studies.

Separate toxicological tests should be performed on toxicologically suspect impurities or minor components present in the test material. If any observed toxicity can be attributed to one of the impurities or minor components, its level should be controlled by use of specifications and manufacturing controls.

6.2.4 Foods from novel sources

Over the past 25 years, a series of developments has made possible the production of foods from unconventional sources (e.g., fungal mycelia and yeast cells) and foods produced by genetic techniques. These foods are intended for consumption, either directly, or after simple physical modification to provide a more acceptable product. They may be consumed in large amounts, even by infants and children, particularly if they are permitted for use as protein supplements in otherwise protein-deficient diets.

Complete chemical identification of such materials may not be feasible, but specifications are necessary to ensure that levels of potentially hazardous contaminants, such as mycotoxins and heavy metals, and other substances of concern, such as nucleic acids, are kept to a minimum. Toxicological evaluations must be closely related to well-defined materials, and evaluations may not be valid for all preparations from the same source material, if different processing methods are used.

When a novel food is intended to replace a significant portion of traditional food in the diet, its likely impact on the nutritional status of consumers requires special consideration. The influence of the introduction of the new substance on the nutrient composition of the diet as a whole should be identified, particularly with respect to groups such as children, the elderly, and "captive populations", e.g., hospital patients and school children. In order not to adversely affect the nutritional quality of the diet, it may be necessary to fortify the substance with vitamins, minerals, or other nutrients.

The nutritional value of the novel food should be assessed initially from its chemical composition with respect to both macronutrients and micronutrients, taking into account the effects of any further processing and storage. The possible influence of components in the novel food, such as antinutritional factors (e.g., inhibitors of enzyme activity or mineral metabolism), on the nutritional value or keeping quality of the remainder of the diet should also be established.

Depending on the nature and intended uses of the novel food, studies in animals may be needed to supplement the chemical studies. If the novel food is intended to be an alternative significant supply of protein, tests on its protein quality will be necessary. In vivo studies will also be needed when it is appropriate to determine (a) the availability of vitamins and minerals in the novel food in comparison with the food it would replace; and (b) any interaction the novel food might have with other items of the diet that would reduce the whole diet's nutritional value. If the novel food is expected to play an important role in the diet, it may be necessary to verify that the results of animal studies can be extrapolated to human beings by measuring the availability of nutrients to human subjects.

In most cases, novel foods constitute a large percentage of the daily diet in animal studies because they are of a non-toxic nature. Therefore, the considerations discussed in section 6.2.3 apply to the toxicological testing and evaluation of foods from novel sources. The sixteenth and seventeenth meetings of the FAO/WHO/UNICEF Protein Advisory Group (PAG) and the United Kingdom Department of Health and Social Security (DHSS) have developed guidelines on the development and testing of foods from novel sources, which should be consulted for a detailed discussion (85 - 87).

With certain classes of substances, such as foods that are modified by recombinant DNA or hybridization techniques to produce what effectively are new cultivars or modified traditional foods, the use of the ADI is not appropriate. However, with many foods produced from novel sources, the use of the ADI is appropriate, because these foods bear little relationship to foods that have been consumed traditionally. The allocation of an ADI is useful to permit the establishment of specifications to ensure microbiological purity and to control chemical contaminants.

After the appropriate animal tests have been performed and a tentative ADI has been established, human volunteer studies to test for specific human effects should be undertaken. The first human study should involve the feeding of a single meal containing the novel food at a known dose level to one volunteer at a time. If no harmful effects are observed with several volunteers, studies involving the feeding of the novel food for a short period (initially about four weeks with follow-up studies of longer duration) should be performed. Different diets incorporating different levels of the novel food should be related to the anticipated levels of human exposure. The closest attention should be paid to matching groups with respect to age, height, weight, sex, alcohol intake, and smoking habits. In addition to having normal control groups, it may be useful to organize studies in which the test groups are fed diets incorporating and not incorporating the novel food in sequential periods, so that each volunteer acts as his own control; blind crossover trials are the most satisfactory. Once it has been determined that the novel food is tolerated well by volunteers at fixed dietary levels, it may be useful to feed it *ad libitum*, for a short period of time, in order to assess its acceptability.

If the novel food is intended for use by a certain community or section of the community (e.g., among a particular ethnic group or by diabetic patients), at least one study should be conducted in the group of people for whom the food is intended.

It may be necessary to conduct allergenicity studies on the novel food because of its composition (e.g., if it is highly proteinaceous) or because the results of animal or human feeding studies suggest that the food might produce hypersensitivity in some people. Important information can be gained by monitoring the health of workers coming into contact with the novel food, such as laboratory staff and employees in the manufacturing plant. In order to detect possible allergenicity of the novel food in the general population, it will generally be essential to monitor a large number of people.

Large-scale acceptability and marketing trials should be undertaken only after the novel food's safety has been demonstrated by the studies indicated above. It may be most useful to restrict the trial to a defined geographical area. The local medical services responsible for the area in which the substance is tested should be alerted so that they may take it into account when evaluating any unusual disease patterns that may appear during or after the test period. Because large numbers of people will be involved in the trials, it may be possible to obtain information about rare food intolerance (e.g., allergic reactions) that may not have been observed in earlier human studies. The extent to which health monitoring should be performed will depend on the nature of the substance and the results of previous toxicological investigations.

REFERENCES

1. FAO/WHO (1984) Evaluation of certain food additives and contaminants. Twenty-eighth report of the Joint FAO/WHO Expert Committee on Food Additives (WHO Technical Report Series No. 710).

2. WHO (1967) Procedures for investigating intentional and unintentional food additives. Report of a WHO Scientific Group, Geneva, World Health Organization (WHO Technical Report Series No. 348).

3. WHO (1974) Assessment of the carcinogenicity and mutagenicity of chemicals. Report of a WHO Scientific Group, Geneva, World Health Organization (WHO Technical Report Series No. 546).

4. FAO/WHO (1961) Evaluation of the carcinogenic hazards of food additives. Fifth report of the Joint FAO/WHO Expert Committee on Food Additives (FAO Nutrition Meetings Report Series No. 29; WHO Technical Report Series No. 220) (out of print).

5. FAO/WHO (1955) Joint FAO/WHO Expert Committee on Nutrition. Fourth report (FAO Nutrition Meetings Report Series No. 9; WHO Technical Report Series No. 97).

6. FAO/WHO (1956) Joint FAO/WHO Conference on Food Additives (FAO Nutrition Meetings Report Series No. 11; WHO Technical Report Series No. 107).

7. FAO/WHO (1963) Second Joint FAO/WHO Conference on Food Additives (FAO Nutrition Meetings Report Series No. 34; WHO Technical Report Series No. 264).

8. FAO/WHO (1974) Report of the Third Joint FAO/WHO Conference on Food Additives and Contaminants (FAO Miscellaneous Meeting Reports Series No. ESN:MMS 74/6; WHO/Food Add./74.43).

9. FAO/WHO (1957) General principles governing the use of food additives. First report of the Joint FAO/WHO Expert Committee on Food Additives (FAO Nutrition Meetings Report Series No. 15; WHO Technical Report Series No. 129).

10. FAO/WHO (1958) Procedures for the testing of intentional food additives to establish their safety for use. Second report of the Joint FAO/WHO Expert Committee on Food

Additives (FAO Nutrition Meetings Report Series No. 17; WHO Technical Report Series No. 144) (out of print).

11. FAO/WHO (1965) Specifications for the identity and purity of food additives and their toxicological evaluation: food colours and some antimicrobials and antioxidants. Eighth Report of the Joint FAO/WHO Expert Committee on Food Additives (FAO Nutrition Meetings Report Series No. 38; WHO Technical Report Series No. 309) (out of print).

12. FAO/WHO (1966) Specifications for the identity and purity of food additives and their toxicological evaluation: some antimicrobials, antioxidants, emulsifiers, stabilizers, flour-treatment agents, acids, and bases. Ninth report of the Joint FAO/WHO Expert Committee on Food Additives (FAO Nutrition Meetings Report Series No. 40; WHO Technical Report Series No. 339) (out of print).

13. WHO (1969) Principles for the testing and evaluation of drugs for carcinogenicity. Report of a WHO Scientific Group, Geneva, World Health Organization (WHO Technical Report Series No. 426).

14. WHO (1971) Evaluation and testing of drugs for mutagencity: principles and problems. Report of a WHO Scientific Group, Geneva, World Health Organization (WHO Technical Report Series No. 482).

15. WHO (1967) Principles for the testing of drugs for teratogencity. Report of a WHO Scientific Group, Geneva, World Health Organization (WHO Technical Report Series No. 364).

16. FAO/WHO (1974) Toxicological evaluation of certain food additives with a review of general principles and of specifications. Seventeenth report of the Joint FAO/WHO Expert Committee on Food Additives (FAO Nutrition Meetings Report Series No. 53; WHO Technical Report Series No. 539 and corrigendum) (out of print).

17. FAO/WHO (1974) Evaluation of certain food additives. Eighteenth report of the Joint FAO/WHO Expert Committee on Food Additives (FAO Nutrition Meetings Report Series No. 54; WHO Technical Report Series No. 557 and corrigendum).

18. FAO/WHO (1975) Evaluation of certain food additives: some food colours, thickening agents, smoke condensates, and certain other substances. Nineteenth report of the Joint FAO/WHO Expert Committee on Food Additives (FAO Nutrition

Meetings Report Series No. 55; WHO Technical Report Series No. 576).

19. FAO/WHO (1976) <u>Evaluation of certain food additives. Twentieth report of the Joint FAO/WHO Expert Committee on Food Additives</u> (FAO Nutrition Meetings Report Series No. 1; WHO Technical Report Series No. 599).

20. FAO/WHO (1978) <u>Evaluation of certain food additives. Twenty-first report of the Joint FAO/WHO Expert Committee on Food Additives</u> (WHO Technical Report Series No. 617).

21. FAO/WHO (1980) <u>Evaluation of certain food additives. Twenty-fourth report of the Joint FAO/WHO Expert Committee on Food Additives</u> (WHO Technical Report Series No. 653).

22. FAO/WHO (1981) <u>Evaluation of certain food additives. Twenty-fifth report of the Joint FAO/WHO Expert Committee on Food Additives</u> (WHO Technical Report Series No. 669).

23. FAO/WHO (1982) <u>Evaluation of certain food additives and contaminants. Twenty-sixth report of the Joint FAO/WHO Expert Committee on Food Additives</u> (WHO Technical Report Series No. 683).

24. FAO/WHO (1983) <u>Evaluation of certain food additives and contaminants. Twenty-seventh report of the Joint FAO/WHO Expert Committee on Food Additives</u> (WHO Technical Report Series No. 696 and corrigenda).

25. FAO/WHO (1962) <u>Specifications for identity and purity of food additives (antimicrobial preservatives and antioxidants). Third report of the Joint FAO/WHO Expert Committee on Food Additives</u>. These specifications were subsequently revised and published as <u>Specifications for identity and purity of food additives. I. Antimicrobial preservatives and antioxidants</u>, Rome, Food and Agriculture Organization of the United Nations (out of print).

26. FAO/WHO (1964) <u>Specifications for the identity and purity of food additives and their toxicological evaluation: emulsifiers, stabilizers, bleaching and maturing agents. Seventh report of the Joint FAO/WHO Expert Committee on Food Additives</u> (FAO Nutrition Meetings Report Series No. 35; WHO Technical Report Series No. 281) (out of print).

27. FAO/WHO (1970) <u>Specifications for the identity and purity of food additives and their toxicological evaluation: some food colours, emulsifiers, stabilizers, anticaking agents,</u>

and certain other substances. Thirteenth report of the Joint FAO/WHO Expert Committee on Food Additives (FAO Nutrition Meetings Report Series No. 46; WHO Technical Report Series No. 445).

28. GEMS (1985) Guidelines for the study of dietary intakes of chemical contaminants, Geneva, World Health Organization (GEMS: Global Environmental Monitoring System) (WHO Offset Publication No. 87).

29. FAO/WHO (1967) Specifications for the identity and purity of food additives and their toxicological evaluation: some emulsifiers and stabilizers and certain other substances. Tenth report of the Joint FAO/WHO Expert Committee on Food Additives (FAO Nutrition Meetings Report Series No. 43; WHO Technical Report Series No. 373).

30. FAO/WHO (1971) Evaluation of food additives: specifications for the identity and purity of food additives and their toxicological evaluation: some extraction solvents and certain other substances; and a review of the technological efficacy of some antimicrobial agents. Fourteenth report of the Joint FAO/WHO Expert Committee on Food Additives (FAO Nutrition Meetings Report Series No. 48; WHO Technical Report Series No. 462).

31. FAO/WHO (1980) Evaluation of certain food additives. Twenty-third report of the Joint FAO/WHO Expert Committee on Food Additives (WHO Technical Report Series No. 648 and corrigenda).

32. FAO/WHO (1978) Evaluation of certain food additives and contaminants. Twenty-second report of the Joint FAO/WHO Expert Committee on Food Additives (WHO Technical Report Series No. 631).

33. FAO/WHO (1985) Evaluation of certain food additives and contaminants. Twenty-ninth report of the Joint FAO/WHO Expert Committee on Food Additives (WHO Technical Report Series No. 733).

34. NCI (1974) An automatic procedure for assessing possible carcinogenic activity of chemicals prior to testing, Palo Alto, California, Stanford Research Institute (prepared for the National Cancer Institute, Contracts N01-CP-33285 and NIH-NCI-71-2045).

35. CRAMER, G.M., FORD, R.A., & HALL, R.L. (1978) Estimation of toxic hazard: a decision tree approach. Food Cosmet. Toxicol., 16: 255-276.

36. FOOD SAFETY COUNCIL (1980) *Proposed system for food safety assessment. Final report of the Scientific Committee of the Food Safety Council*, Washington DC, Food Safety Council.

37. FDA (1982) *Toxicological principles for the safety assessment of direct food additives and color additives used in food*, Washington DC, Bureau of Foods, US Food and Drug Adminstration.

38. DHSS (1982) *Guidelines for the testing of chemicals for toxicity*, London, Department of Health and Social Security, Committee on Toxicity of Chemicals in Food, Consumer Products and the Environment (Report on Health and Social Subjects No. 27).

39. FAO/WHO (1962) *Evaluation of the toxicity of a number of antimicrobials and antioxidants. Sixth report of the Joint FAO/WHO Expert Committee on Food Additives* (FAO Nutrition Meetings Report Series No. 31; WHO Technical Report Series No. 228) (out of print).

40. VETTORAZZI, G., ed. (1980) *Handbook of international food regulatory toxicology. I. Evaluations*, New York, SP Medical and Scientific Books.

41. FAO/WHO (1968) *Specifications for the identity and purity of food additives and their toxicological evaluation: some flavouring substances and non-nutritive sweetening agents. Eleventh report of the Joint FAO/WHO Expert Committee on Food Additives* (FAO Nutrition Meetings Report Series No. 44; WHO Technical Report Series No. 383).

42. FAO/WHO (1972) *Evaluation of food additives: some enzymes, modified starches, and certain other substances: toxicological evaluations and specifications and a review of the technological efficacy of some antioxidants. Fifteenth report of the Joint FAO/WHO Expert Committee on Food Additives* (FAO Nutrition Meetings Report Series No. 50; WHO Technical Report Series No. 488).

43. FAO (1983) *Guide to specifications: general notices, general methods, identification tests, test solutions, and other reference materials*, Rome, Food and Agriculture Organization of the United Nations (FAO Food and Nutrition Paper No. 5, Rev. 1).

44. FAO (1978) *Guide to specifications: general notices, general methods, identification tests, test solutions, and other reference materials*, Rome, Food and Agriculture

Organization of the United Nations (FAO Food and Nutrition Paper No. 5).

45. IARC (1983) <u>Approaches to classifying chemical carcinogens according to mechanism of action. Joint IARC/IPCS/CEC Working Group report</u>, Lyons, International Agency for Research on Cancer (IARC Internal Technical Report No. 83/001).

46. FEDERAL REGISTER (1985) Chemical carcinogens: a review of the science and its associated principles. <u>Fed. Reg.</u>, 50: 10371-10442 (March 14).

47. FOX, T.R. & WATANABE, P.G. (1985) Detection of a cellular oncogene in spontaneous liver tumors of B6C3F1 mice. <u>Science</u>, <u>228</u>: 596-597.

48. RIBELIN, W.E., ROLOFF, M.V., & HOUSER, R.M. (1984) Minimally functional rat adrenal medullary pheochromocytomas. <u>Vet. Pathol.</u>, <u>21</u>: 281-285.

49. ROE, F.J.C. & BAR, A. (1985) Enzootic and epizootic adrenal medullary proliferative disease of rats: influence of dietary factors which affect calcium absorption. <u>Human Toxicol.</u>, <u>4</u>: 27-52.

50. NTP (1984) <u>Report of the NTP Ad Hoc Panel on Chemical Carcinogenesis Testing and Evaluation</u>, Washington DC, Board of Scientific Counselors, National Toxicology Program, US Department of Health and Human Services.

51. BAEDER, C., WICKRAMARATNE, G.A.S., HUMMLER, H., MERKLE, J., SCHON, H., & TUCHMANN-DUPLESSIS, H. (1985) Identification and assessment of the effects of chemicals on reproduction and development (reproductive toxicology). <u>Food Chem. Toxicol.</u>, <u>23</u>: 377-388.

52. KOCHENOUR, N.K. (1984) Adverse pregnancy outcome: sensitive periods, types of adverse outcomes, and relationships with critical exposure periods. In: <u>The new frontier in occupational and environmental health research</u>, New York, Alan R. Liss, pp. 229-235.

53. WHO (1984) <u>EHC 30: Principles for evaluating health risks to progeny associated with exposure to chemicals during pregnancy</u>, Geneva, World Health Organization, 123 pp.

54. VOUK, V.B. & SHEEHAN, P.J., ed. (1983) <u>Methods for assessing the effects of chemicals on reproductive functions</u>, New York, John Wiley and Sons.

55. WILSON, J.G. (1973) Environment and birth defects, New York, London, Academic Press.

56. JOHNSON, E.M. & CHRISTIAN, M.S. (1984) When is a teratology study not an evaluation of teratogenicity? J. Am. Coll. Toxicol., 3: 431-434.

57. WHO (1985) EHC 47: Summary report on the evaluation of short-term tests for carcinogens (Collaborative study on in vitro tests), Geneva, World Health Organization, 74 pp (In: Ashby, J., de Serres, F.J., Draper, M., Ishidate, M., Jr, Margolin, B.H., Matter, B.E., & Shelby, M.D., ed. Progress in mutation research, Amsterdam, Elsevier, Vol. 5).

58. NEUBERT, D. (1982) The use of culture techniques in studies on prenatal toxicity. Pharmacol. Ther., 18: 397-434.

59. ROSLAND, I.R. & WALKER, R. (1983) The gastrointestinal tract in food toxicology. In: Conning, D.M., & Lansdown, A.B.G., ed. Toxic hazards in food, London, Croom Helm, pp. 183-274.

60. FAO/WHO (1972) Evaluation of certain food additives and the contaminants mercury, lead, and cadmium. Sixteenth report of the Joint FAO/WHO Expert Committee on Food Additives (FAO Nutrition Meetings Report Series No. 51; WHO Technical Report Series No. 505 and corrigendum).

61. WHO (1986) EHC 59: Principles for evaluating health risks from chemicals during infancy and early childhood: the need for a special approach, Geneva, World Health Organization, 73 pp.

62. TOMATIS, L., MOHR, U., & DAVIS, W., ed. (1973) Transplacental carcinogenesis, Lyons, International Agency for Research on Cancer (IARC Scientific Publication No. 4) (Proceedings of a Meeting held at the Medizinische Hochschule, Hanover, Federal Republic of Germany, 6-7 October, 1971).

63. ALTHOFF, J. & GRANDJEAN, C. (1979) In vivo studies in Syrian golden hamsters: a transplacental bioassay of ten nitrosamines. Natl Cancer Inst. Monogr. Ser., 51: 251-255.

64. HARADA, Y. (1968) In: Kutsana, M., ed. Minamata disease, Kumamoto, Japan, Kumamoto University Press, pp. 93-117.

65. BOWMAN, R.E., HEIRONIMUS, M.P., & BARSOTTI, D.A. (1981) Locomotor hyperactivity in PCB-exposed rhesus monkeys. Neurotoxicity, 2: 251-268.

66. ROGAN, W.J., BAGNIEWSKA, A., & DAMSTRA, T. (1980) Pollutants in breast milk. N. Engl. J. Med., 302: 1450-1453.

67. LAMARTINIERE, C.A., LUTHER, M.A., LUCIER, G.W., & ILLSEY, N.P. (1982) Altered imprinting of rat liver monoamine oxidase by o,p'-DDT and methoxychlor. Biochem. Pharmacol., 31: 647-651.

68. KATO, R. & TAKANAKA, A. (1968) Effect of phenobarbital on electron transport system, oxidation and reduction of drugs in liver microsomes of rats of different ages. J. Biochem., 63: 406-408.

69. KATO, R. & TAKANAKA, A. (1968) Metabolism of drugs in old rats. I. Activities of NADPH-linked electron transport and drug-metabolizing enzyme systems in liver microsomes in old rats. Jpn. J. Pharmacol., 18: 381-388.

70. KATO, R. & TAKANAKA, A. (1968) Metabolism of drugs in old rats. II. Metabolism in vivo and effects of drugs in old rats. Jpn. J. Pharmacol., 18: 389-396.

71. SCHMUCKER, D.L. & WANG, R.K. (1980) Age-related changes in liver drug-metabolizing enzymes. Exp. Gerontol., 15: 321-329.

72. TANNENBAUM, A. & SILVERSTONE, H. (1953) Nutrition in relation to cancer. Advances Cancer Res., 1: 451-501.

73. ROSS, M.H., BRAS, G., & RAGBEER, M.S. (1970) Influence of protein and caloric intake upon spontaneous tumour incidence of the anterior pituitary gland of the rat. J. Nutr., 100: 177-189.

74. FAO/WHO (1981) Toxicological evaluation of certain food additives (WHO Food Additives Series No. 16).

75. GRICE, H.C., ed. (1984) The selection of doses in chronic toxicity/carcinogenicity studies and age-associated (geriatric) pathology: its impact on long-term toxicity studies, New York, Springer-Verlag (Current Issues in Toxicology. Sponsored by the International Life Sciences Institute).

76. WHO (1978) EHC 6: Principles and methods for evaluating the toxicity of chemicals. Part I, Geneva, World Health Organization, 264 pp.

77. PAGET, G.E. (1970) The design and interpretation of toxicity tests. Methods Toxicol., 1970: 1-10.

78. MORGAN, R.W. & WONG, O. (1985) A review of epidemiological studies on artificial sweeteners and bladder cancer. Food Chem. Toxicol., 23: 529-533.

79. ATKINS, F.M. & METCALFE, D.D. (1984) The diagnosis and treatment of food allergy. Ann. Rev. Nutr., 20: 233-255.

80. BOCK, S.A. & MAY, C.D. (1983) Adverse reactions to food caused by sensitivity. In: Middleton, E., Jr, Reed, C.E., & Ellis, E.F., ed. Allergy principles and practice, 2nd ed., St. Louis, Missouri, C.V. Mosby Company, p. 1415.

81. SIRAGANIAN, R.P. & HOOK, W.A. (1980) Histamine release and assay methods for the study of human allergy. In: Rose, N.R. & Friedman, H., ed. Manual of clinical immunology, 2nd ed., Washington DC, American Society of Microbiology, p. 808.

82. SANTRACH, P.J., PARKER, J.L., JONES, R.T., & YUNGINGER, J.W. (1981) Diagnostic and therapeutic applications of a modified RAST and comparison with the conventional RAST. J. Allergy clin. Immunol., 67: 97.

83. BOCK, S.A., LEE, W.Y., REMIGO, L.K., & MAY, C.D. (1978) Studies of hypersensitivity reactions to foods in infants and children. J. Allergy clin. Immunol., 62: 327.

84. SAMPSON, H.A. & ALBERGO, R. (1984) Comparison of results of skin tests, RAST, and double-blind, placebo-controlled food challenges in children with atopic dermatitis. J. clin. Immunol., 74: 26.

85. Protein advisory group report on the FAO/WHO/UNICEF protein advisory group 16th meeting, Geneva, 8-11 September, 1969 (Unpublished PAG Meeting Report Document 3.14/5).

86. Protein advisory group report on the FAO/WHO/UNICEF protein advisory group 17th meeting, New York, 25-28 May, 1970 (Unpublished PAG Meeting Report Document 3.14/8).

87. Memorandum on the testing of novel foods, incorporating guidelines for testing by the advisory committee on irradiated and novel foods, Department of Health and Social

Security, Ministry of Agriculture, Fisheries and Food, Scottish Home and Health Department, Welsh Office, and Department of Health and Social Services, Northern Ireland, December, 1984.

ANNEX I. GLOSSARY

I.1 Abbreviations Used in this Document

ADI:	Acceptable Daily Intake (see definition)
CCFA:	Codex Committee on Food Additives (see definition of Codex Alimentarius Commission)
COT:	Committee on Toxicity (United Kingdom)
CSM:	Committee on Safety of Medicines (United Kingdom)
EEC:	European Economic Community
EPA:	Environmental Protection Agency (USA)
FAO:	Food and Agriculture Organization of the United Nations
FDA:	Food and Drug Administration (USA)
FSC:	Food Safety Council, Washington, DC, USA
GEMS:	Global Environmental Monitoring System
GLP:	Good Laboratory Practice
IARC:	International Agency for Research on Cancer
IPCS	International Programme on Chemical Safety
JECFA:	Joint FAO/WHO Expert Committee on Food Additives (see definition)
LD_{50}	Lethal Dose, median
MAFF:	Ministry of Agriculture, Forestry and Fisheries (Japan)
MHW:	Ministry of Health and Welfare (Japan)
MTD:	Maximum Tolerated Dose (see definition)
OECD:	Organisation for Economic Cooperation and Development
PMTDI:	Provisional Maximum Tolerable Daily Intake (see definition)

PSPS: Pesticides Safety Precautions Scheme (United Kingdom)

PTWI: Provisional Tolerable Weekly Intake (see definition)

SCOPE: Scientific Committee on Problems of the Environment of the International Council of Scientific Unions

UNICEF: United Nations Childrens' Fund

WHO: World Health Organization

I.2 Definitions of Terms Used in this Document

Acceptable daily intake: An estimate by JECFA of the amount of a food additive, expressed on a body weight basis, that can be ingested daily over a lifetime without appreciable health risk (standard man = 60 kg).

Acceptable daily intake not allocated: See no ADI allocated.

Acceptable daily intake not specified: A term applicable to a food substance of very low toxicity which, on the basis of the available data (chemical, biochemical, toxicological, and other), the total dietary intake of the substance arising from its use at the levels necessary to achieve the desired effect and from its acceptable background in food does not, in the opinion of JECFA, represent a hazard to health. For that reason, and for reasons stated in individual evaluations, the establishment of an acceptable daily intake expressed in numerical form is not deemed necessary. An additive meeting this criterion must be used within the bounds of good manufacturing practice, i.e., it should be technologically efficacious and should be used at the lowest level necessary to achieve this effect, it should not conceal inferior food quality or adulteration, and it should not create a nutritional imbalance.

Codex Alimentarius Commission: The Commission was formed in 1962 to implement the Joint FAO/WHO Food Standards Programme. The Commission is an intergovernmental body made up of more than 120 Member Nations, the delegates of whom represent their own countries. The Commission's work of harmonizing food standards is carried out through various committees, one of which is the Codex Committee on Food Additives (CCFA). JECFA serves as the advisory body to the Codex Alimentarius Commission on all scientific matters concerning food additives.

Conceptus: All products of conception derived from and including the fertilized ovum at any time during pregnancy, including the embryo or fetus and embryonic membranes.

Developmental toxicity: Any adverse effects induced prior to attainment of adult life, including effects induced or manifested in the embryonic or fetal period and those induced or manifested postnatally (before sexual maturity).

Effect: A biological change in an organism, organ, or tissue.

Elimination (in metabolism): The expelling of a substance or other material from the body (or a defined part thereof), usually by a process of extrusion or exclusion, but sometimes through metabolic transformation.

Embryo/fetotoxicity: Any toxic effect on the conceptus resulting from prenatal exposure, including structural or functional abnormalities or postnatal manifestation of such effects.

Embryonic period: The period from conception to the end of major organogenesis. Generally, the organ systems are identifiable at the end of this period.

Enterohepatic circulation: Intestinal reabsorption of material that has been excreted through the bile followed by transfer back to the liver, making it available for biliary excretion again.

Fetal period: The period from the end of embryogenesis to the completion of pregnancy.

Food allergy: A form of food intolerance in which there is evidence of an abnormal immunological reaction to the food. "Immediate allergic reactions" are those which occur within minutes to hours after ingestion of the offending food, while reactions beginning several hours to days after food exposure are characterized as "delayed allergic reactions".

Food intolerance: A reproducible, unpleasant reaction to a food or food ingredient, including reactions due to immunological effects, biochemical factors such as enzyme deficiencies, and anaphylactoid reactions that often include histamine release.

Group acceptable daily intake: An acceptable daily intake established for a group of compounds that display similar toxic effects, thus limiting their cumulative intake.

Irreducible level (of a food contaminant): That concentration of a substance which cannot be eliminated from a food without involving the discarding of that food altogether, severely compromising the ultimate availability of major food supplies.

JECFA: JECFA is a technical committee of specialists acting in their individual capacities. Each JECFA is a separately-constituted committee, and when either the term "JECFA" or "the Committee" is used, it is meant to imply the common policy or combined output of the separate meetings over the years.

Long-term toxicity study: A study in which animals are observed during the whole life span (or the major part of the life span) and in which exposure to the test material takes place over the whole observation time or a substantial part thereof. The term chronic toxicity study is used sometimes as a synonym for "long-term toxicity study".

Maximum tolerated dose: A term in common use in carcinogenicity testing meaning a dose that does not shorten life expectancy nor produce signs of toxicity other than those due to cancer (operationally, the MTD has been set as the maximum dose level at which a substance induces a decrement in weight gain of no greater than 10% in a subchronic toxicity test).

No ADI allocated: Terminology used by JECFA in situations where an ADI is not established for a substance under consideration because (a) insufficient safety information is available; (b) no information is available on its food use; or (c) specifications for identity and purity have not been developed.

No-observed-effect level: The greatest concentration or amount of an agent, found by study or observation, that causes no detectable, usually adverse, alteration of morphology, functional capacity, growth, development, or lifespan of the target.

Novel food: A food or food ingredient produced from raw materials not normally used for human consumption or food that is severely modified by the introduction of new processes not previously used in the production of food.

Processing aid: A substance added to food during processing, but subsequently removed. Traces of a processing aid may remain with the food.

Provisional maximum tolerable daily intake: The end-point used by JECFA for contaminants with no cumulative properties. Its value represents permissible human exposure as a result of the natural occurrence of the substance in food and in drinking water. In the case of trace elements that are both essential nutrients and unavoidable constituents of food, a range is expressed, the lower value representing the level of essentiality and the upper value the PMTDI.

Provisional tolerable weekly intake: The end-point used by JECFA for food contaminants such as heavy metals with cumulative properties. Its value represents permissible human weekly exposure to those contaminants unavoidably associated with the consumption of otherwise wholesome and nutritious foods.

Reproductive effects: To test for the effects of exposure to low levels of chemicals exceeding the life span of one generation, tests have been developed covering several reproductive cycles. In the three-generation test, the animals are exposed through three complete reproductive cycles (starting with the F_0 generation at weaning). These tests, which include exposure *in utero* and through the milk, have been used in particular for assessing toxic effects related to reproduction.

Safety factor: A factor applied by JECFA to the no-observed-effect level to derive an acceptable daily intake (the no-observed-effect level is divided by the safety factor to calculate the ADI). The value of the safety factor depends on the nature of the toxic effect, the size and type of population to be protected, and the quality of the toxicological information available.

Short-term toxicity study: An animal study (sometimes called a subacute or subchronic study) in which the effects produced by the test material, when administered in repeated doses (or continuously in food or drinking-water) over a period of about 90 days, are studied.

Temporary acceptable daily intake: Used by JECFA when data are sufficient to conclude that use of the substance is safe over the relatively short period of time required to generate and evaluate further safety data, but are insufficient to conclude that use of the substance is safe over a lifetime. A higher-than-normal safety factor is used when establishing a temporary ADI and an expiration date is established by which time appropriate data to resolve the safety issue should be submitted to JECFA.

Teratogen: An agent which, when administered prenatally, induces permanent abnormalities in structure.

Teratogenicity: The property (or potential) to produce structural malformations or defects in an embryo or fetus.

Transplacental carcinogenesis: The appearance of neoplasia in the progeny of females exposed to chemical agents during pregnancy.

ANNEX II. STATISTICAL ASPECTS OF TOXICITY STUDIES

II.1 Summary

Statistical design and analysis should aim at eliminating sources of potential bias and minimizing the role of chance. The application of these principles in the experimental design and conduct of toxicological studies is discussed under 10 headings: choice of species, dose levels, number of animals, duration of the study, accuracy of determinations, stratification, randomization, adequacy of control groups, animal placement, and data recording. A number of general considerations to be borne in mind when conducting statistical analyses are also discussed: experimental and observational units, types of response variable, types of between-group comparisons, stratification, age adjustment, multiple observations per animal, hypothesis testing and probability values, and multiple comparisons. Finally, some recommended methods of statistical analysis are summarized.

II.2 Introduction

These guidelines are intended primarily to provide the experimental scientist without statistical qualifications with some insight into statistical aspects of toxicological studies. Considerations relating to the design and conduct of studies and to the analysis and interpretation of results are discussed, emphasizing the principles involved rather than the mathematical details. While the experimentalist should have sufficient information to deal with many standard situations, the need for the advice of an expert statistician, when dealing with toxicological data, cannot be overemphasized. Scientific journals frequently contain papers describing studies in which the conclusions of the author(s) cannot be supported because of deficiencies in statistical methodology, which could have been avoided had the advice of a qualified statistician been available to the researcher.

II.3 Sources of Difference Between Treated and Control Groups

An objective of many toxicity studies is to determine whether a treatment elicits a response. However, the observation of a difference in response between a treated and a control group does not necessarily mean that the difference is a result of the treatment. There are two other potential causes of difference, bias and chance.

Bias implies systematic differences other than treatment between the groups, in other words, failure to compare like with like. Properly conducted studies analysed appropriately can eliminate bias.

Chance factors cannot be wholly excluded, because identically-treated animals will not all respond identically, however carefully the study is conducted. While it is impossible to be absolutely certain that even very extreme differences in response are not due to chance, appropriate statistical analysis will allow the experimentalist to assess the probability of a "false positive", that is, the probability of the observed difference having occurred had there been no effect of treatment at all. The smaller the probability, the greater the confidence of having found a real effect. To improve the likelihood of detecting a true effect with confidence, it is necessary to try to minimize the role of chance by seeking to ensure that the "signal" can be recognized above the "noise".

II.4 Experimental Design and Conduct

Ten aspects of the experimental design and conduct of toxicity studies are considered below, the first six being involved primarily with minimizing the role of chance and the last four being particularly relevant to the avoidance of bias. For convenience, the principles are illustrated with reference to a long-term carcinogenicity study.

II.4.1 Choice of species

While maximizing the "signal" means avoiding a species in which the response of interest is very rare, the use of an overresponsive species also has problems. Thus, to achieve the same level of statistical significance in comparing a treated group with a 5% response and a control group with a 0% response requires only one tenth as many animals as when the responses are 55 and 50%, respectively. Furthermore, it is not certain that an increased incidence of a lesion that is a common spontaneous finding in the animal species used (such as pituitary tumours in Wistar rats) provides biological evidence of an effect that can be extrapolated to other species. Other considerations related to the choice of species, whether they be practical (short life span, small size, availability, existence of detailed knowledge of the species) or more theoretical (biochemical, physiological, or anatomical similarity to man), do not really pose statistical problems.

II.4.2 Dose levels

Dose selection is an important and controversial element in the development of a protocol for a toxicity bioassay. On biological grounds, it would be ideal to test only at dose levels comparable with those to which human beings are exposed. On statistical and economic grounds, this is not usually practicable because the effect will be too small to detect without very large numbers of animals. To avoid the possibility of missing an effect that would occur in a small proportion of millions of exposed human beings in a study on hundreds or even thousands of animals, it is normally appropriate to test animals at dose levels many times higher than the maximum human exposure level. Then, assuming any effect that exists is dose-related, the demonstration of a non-significant increase in response at a high dose level, though not providing evidence of absolute safety (an impossible goal), can give reasonable grounds for believing that any effects that might occur at a very much lower dose level would be, at most, very slight.

A particular problem with this procedure is to decide how high the dose level should be. In long-term carcinogenicity studies, the dose should clearly be one that is not so great that the animals die from toxic effects before they have a chance to get cancer. On the basis of these principles, the International Agency for Research on Cancer (1) has recommended that the high dose should be one expected on the basis of an adequate short-term study to produce some toxicity when administered for the duration of the study, but should not induce: (a) overt toxicity, i.e., appreciable death of cells or organ disfunction, as determined by appropriate methods; (b) toxic manifestations that are predicted materially to reduce the life span of the animals, except as a result of the development of neoplasms; or (c) 10% or greater retardation of body-weight gain compared with that in control animals.

If the substance seems completely non-toxic, the high dose may represent about 5% of the diet, or even more for substances such as some nutritive food ingredients.

It is important to have more than one dose level for a number of reasons. One is to compensate for the possibility that a misjudgment has occurred and that the highest dose may prove to be toxic. A second is that the metabolic pathways may differ at the various dose levels. A third reason is that the whole point of the study may be to obtain dose-response information. Finally, it may be necessary to ensure that an effect does not occur at dose levels in the range to be used by man.

II.4.3 Number of animals

The number of animals to be used is clearly an important determinant of the precision of the findings. The calculation of the appropriate number depends on:

(a) The critical difference, i.e., the size of the effect to be detected;

(b) the false positive rate, i.e., the probability of an effect being detected when none exists (known as the "type I error" or the " α level"); and

(c) the false negative rate, i.e., the probability of no effect being detected when one of exactly the critical size exists (known as the "type II error" or the " ß level").

A reduction in any of these factors means an increase in the number of animals required.

The method of calculation of the number depends on the experimental design and the type of statistical analysis envisaged. Tables are available for a number of standard situations. To give an idea how the numbers depend on the critical difference and on the α and β levels, Tables 1 and 2 give examples of two common situations, both of which are related to a study in which there is a control and a treated group. The first is related to a continuous variable assumed to be normally distributed, with the critical difference expressed in terms of the number of standard deviations (δ) by which the treated group differs from the control group. Thus, given a control response known from past experience to have a mean value of 50 units with a standard deviation of 20 units, two groups of 36 animals each would be needed to have a 90% chance ($\beta = 0.10$) of detecting a difference in response of 10 units ($\delta = 10/20 = 0.5$) at the 95% confidence level ($\alpha = 0.05$).

In the second situation, two proportions are compared. Here the numbers of animals depend not only on the ratio of proportions, but also on the assumed proportion in the controls. Thus, when the control response is expected to be 10%, the numbers of animals required in each group to detect an increased response by a factor of 1.5 (r) is 920, assuming again an α level of 0.05 and a β level of 0.1, whereas if the control response is expected to be 50%, the numbers required would be 79 per group.

For more complex situations, the advice of a professional statistician should be sought, though a general rule is that to increase precision (i.e., decrease the size of the critical difference) by a factor n, the number of animals required will have to be increased by a factor of approximately n squared.

Table 1. Number of animals required in each of a control and treated group in order to have a probability $(1 - \beta)$ of picking up a difference of δ standard deviations as significant at the 100 $(1 - \alpha)$ percent confidence level for a normally distributed variable

Single-side test[a]									
Double-sided test[a]	$\alpha = 0.005$			$\alpha = 0.025$			$\alpha = 0.05$		
	$\alpha = 0.01$			$\alpha = 0.05$			$\alpha = 0.1$		
$\beta =$	0.01	0.1	0.5	0.01	0.1	0.5	0.01	0.1	0.5
$\delta = $ 0.5	100	63	30	76	44	18	65	36	13
0.75	47	30	16	35	21	9	30	17	7
1.0	28	19	10	21	13	6	18	11	5
1.5	15	11	7	11	7		9	6	
2.0	10	8	5	7	5		6		

[a] See section II.5.7 for definitions of single-sided and double-sided tests.

Table 2. Number of animals required in each of a control and treated group in order to have a probability $(1 - \beta)$ of picking up a proportional increase by a factor r as significant at the 100 $(1 - \alpha)$ percent confidence level for a binomially distributed variable

Single-side test[a]									
Double-sided test[a]	$\alpha = 0.005$			$\alpha = 0.025$			$\alpha = 0.05$		
	$\alpha = 0.01$			$\alpha = 0.05$			$\alpha = 0.1$		
$\beta =$	0.01	0.1	0.5	0.01	0.1	0.5	0.01	0.1	0.5
Control level = 10%									
$r =$ 1.25	7679	4754	2120	5871	3358	1228	5039	2737	865
1.5	2103	1302	581	1608	920	337	1380	750	237
2.0	613	380	170	469	268	98	403	219	69
Control level = 20%									
$r =$ 1.25	3353	2076	926	2563	1466	536	2200	1195	378
1.5	902	558	249	689	395	145	592	322	102
2.0	253	157	70	193	111	41	166	90	29
Control level = 50%									
$r =$ 1.25	757	469	209	579	331	122	497	270	86
1.5	181	112	50	138	79	29	119	65	21
2.0	37	23	10	28	16	6	24	13	5

[a] See section II.5.7 for definitions of single-sided and double-sided tests.

When a number of treatments are to be tested in a study, each to be compared with a single untreated control group, it is advisable that more animals be included in the control group than in each of the treated groups, because the precision of the control group results is relatively more important. When all groups are of equal interest, it is appropriate to have approximately the square root of k times as many animals in the control group as in each of the k treated groups.

One point frequently misunderstood by the experimental scientist is related to the number of animals required in studies in which more than one treatment is investigated in a crossed design. If, for example, compounds A and B are being compared, and each is dissolved in two different solvents, in a 2 x 2 design with 4 groups, calculations of sample size to gain an overall verdict on the difference between the two compounds should generally be based on the overall numbers of animals treated with each compound for both solvents combined, unless there is reason to expect compound-solvent interaction, i.e., that the compound A/compound B difference depends on which solvent is used. Conversely, if it has been decided that 2 groups of 100 animals each are sufficient for attaining a given level of precision concerning the differences in effects of a treatment, additional information and another factor (or factors) of interest can be obtained without requiring any additional animals.

II.4.4 Duration of the study

The duration of the study can also markedly affect the sensitivity of tests. This is particularly so in long-term carcinogenicity studies in which the great majority of cancers are seen in the latter half of an animal's lifetime. Thus, while studies should not be terminated too early, it is also important that they do not go on too long. This is because the last few weeks or months may produce relatively little data at a disproportionate cost, and diseases of extreme old age may be of little interest in themselves but may render it more difficult to detect tumours and other conditions that are of interest. Where the study is of the prevalence of an age-related non-lethal condition observable only at death that ultimately occurs in all or nearly all of the animals, early termination is required. In this situation the greatest sensitivity is obtained when the average prevalence is about 50%.

II.4.5 Accuracy of determinations

Accuracy of observations is clearly important in minimizing error. The advent of good laboratory practice and quality control units has done much to improve the quality of recording observations, but the quality of the study still depends on interested and diligent personnel.

II.4.6 Stratification

To detect a treatment difference with accuracy, the groups being compared should be as homogeneous as possible with respect to other known causes of the response of interest. Consider, for example, a set of animals thought to be homogenous (but which, in fact, consist of two genetically different substrains) in which the following measurements of body weight were obtained in groups of 10 treated and 10 control animals, the underlined readings relating to the first of the two substrains:

Control: <u>181 192 217</u> 290 321 292 307 347 276 256

Treated: <u>222 249 232 284 270 215 265</u> 378 328 391.

If the substrain is ignored, the variability of the data increases so that more controls are required to detect a treatment effect and, if unequal numbers of each substrain are present in each group, may bias the comparison. In the example given the means are as follows:

	Substrain 1	Substrain 2	Total
Control	196.7	298.4	267.9
Treated	248.1	365.7	283.4
Difference	51.5	67.2	15.5

Although in each substrain the treatment results in an increase in body weight of over 50 units, the greater number of the lower-weight strain in the treated groups means that the difference observed is much less.

There are two ways to take account of the substrain difference and to achieve a more precise answer. One is to use substrain as a "stratifying variable" at the analysis stage. This involves carrying out separate analyses at each level of the variable considered and combining the results for an overall conclusion about the treatment effect. However, it does not preclude the possibility that the proportions of each substrain in each group are so different that the data provide substantially less comparative information than might otherwise be achieved. In the extreme case, if all control animals were of substrain 1 and all treated animals were of substrain 2, the study would be worthless to determine whether differences were due to treatment or substrain. To obviate this possibility, substrain can be used as a "blocking factor" in the design. In this case, animals in each substrain are allocated equally to control and treated groups. Although this removes bias, it is still necessary to treat strain as a stratifying variable in statistical analysis to increase precision.

When more than one known factor affects the response, all can be taken into account simultaneously. Both at the design stage, or retrospectively in the analysis, the results are treated at each combination of levels of the factors. Thus, to block for substrain, sex, and room where 3 experimental rooms were needed to house the animals, 12 mini-studies, one for each of the 2 (substrains) x 2 (sexes) x 3 (rooms) combinations would be set up.

II.4.7 Randomization

Random allocation, or randomization, of animals in treatment groups is an essential of good experimental design. If not carried out, it is not strictly possible to tell whether a difference between groups is a result of differences in the treatment applied or is due to some other relevant factor. A fundamental on which statistical methodology is based is that the probability of a particular response occurring is equal for each animal, regardless of group. The ability to randomize easily is a major advantage that animal studies have over epidemiological studies.

The process of randomization eliminates bias, so that statistical analysis is concerned only with assessing the probability of an observed difference happening by chance. The smaller the probability, the more it suggests a true treatment effect. The procedure used for randomization should genuinely ensure that all possible assignments of animals to treatment groups are equally probable. Such equal probabilities are best achieved with pseudorandom numbers, as found in tables or produced by computer, it being difficult to ensure that apparently random devices such as dice or playing cards really are random. Randomization should never be based on a system of testing animals haphazardly, as they come, and assigning them to successive treatment groups. Not only do human beings find it virtually impossible to generate random sequences unaided, but it is well known that the first animals selected may differ markedly from the last, who are more active and avoid being caught.

In many experimental situations it is adequate to randomly allocate all the animals to treatment groups, but, in some, the technique of stratified random sampling is preferred. In this technique, the animals are first divided into subgroups ("strata"), according to factors known or believed to be strongly related to the response, with random allocation to treatment groups then being carried out within each stratum. Sex is normally treated as a stratifying variable. In a large study in which animals are delivered in batches, batch could also be treated in this way, each batch forming a smaller study, the results from which can be combined in the analysis.

The above discussion on randomization and stratification has been concerned primarily with the allocation of animals to treatment groups. The same principles apply to anything that can affect the recorded response. Thus, in a two-group study, measurements of some biochemical parameter should not be made for the first group in the morning and for the second group in the afternoon. While the major part of such potential bias can be averted fairly easily by various simple procedures, such as doing alternate measurements on treated and control animals, randomization is preferable. Although many different procedures throughout a study (feeding, weighing, observation, clinical chemistry, and pathological examinations) require consideration in this way, the same random number can usually be applied to all the procedures. Thus, if the cage position of the animals is randomly allocated and does not depend on treatment, the animals can always be handled in the same cage sequence.

II.4.8 Adequacy of control groups

The principle of comparing like with like implies that control groups should be randomly allocated from the same control source as the treatment groups. While historical control data can be of value in the interpretation of rarer findings in treated animals, there is so much evidence of quite large systematic differences in response between apparently identical untreated control groups tested at different times that it is often impossible to be sure whether a difference seen between a treated group and a historic group is really due to treatment at all.

It is also essential to be sure that the treated group differs from the control group only with respect to the treatment of interest. Thus, if a treatment is applied in a solvent, an untreated control is not a proper basis for comparison, as one cannot be sure whether observed differences are a result of the treatment or the solvent. In this case, the appropriate control group would be one in which animals are given only the solvent.

II.4.9 Animal placement

The general underlying requirement to avoid systematic differences between groups other than their treatment also demands that attention be given to the question of animal placement. If all treated animals are placed on the highest racks or are at one end of the room, differences in heating, lighting, or ventilation may produce effects that are erroneously attributed to treatment. Such systematic differences should be avoided, and, in many cases, randomization of cage positions is desirable. This may not be possible in some circumstances, such as with studies involving volatile agents where cross-contamination can occur.

II.4.10 Data recording

The application of the principle of comparing like with like means the avoidance of systematic bias in data recording practices. Two distinctly different types of bias can occur. The first is a systematic shift in the standard of measurement with time, coupled with a tendency for the time of measurements to vary from treatment to treatment. The second is that awareness of the treatment may affect the values recorded by the measurer, consciously or subconsciously. The second bias is circumvented by the animals' treatment not being known to the measurer, i.e., the readings being carried out "blind". Although not always practical (that an animal is treated may be obvious from its appearance), laboratories should organize their data recording practices so that, at least for subjective measurements, the observations are made "blind".

The problem of avoidance of bias due to differences in time of observation is a particularly important one in histopathological assessment, especially for the recording of lesions of a graded severity, and in large studies in which the slides may take the pathologist more than a year to read. When more than one pathologist reads the slides, there should be discussion between them as to standardization of terminology and data to be recorded, and each should read a random or a stratified set of the slides to avoid bias.

II.5 <u>Statistical Analysis - General Considerations</u>

In the simplest situation, animals are randomly assigned to a treated, or a control, group and one observation is made on each animal, the objective of the statistical analysis being to determine whether the distribution of responses in the treated group differs from that in the control group. Before summarizing some of the appropriate techniques for analysis, a number of more general points underlying the choice of the correct method and interpretation of the results will be discussed.

II.5.1 Experimental and observational units

In the simple example cited above, the animal is both the "experimental unit" and the "observational unit". This is not always so. In the case of feeding studies, the cage, rather than the animal, is usually the experimental unit in that it is the cage, rather than the animal, that is assigned to the treatment. In the case of histopathology, observations are often made from multiple sections per animal in which case the section rather than the animal is the observational unit. For the purpose of determining treatment effects by the methods described below, it is important that each experimental unit provides only one item of data for analysis, as the methods are all based on the assumption that individual data items are statistically independent. If multiple observations per experimental unit are made, these observations should be combined in some suitable way into an overall observation for that experimental unit before analysis. Thus, in a study in which 20 animals were assigned to two treatment groups of 10 animals each and in which measurements of the weight of both kidneys were made individually, it would be wrong to carry out an analysis in which the 20 weights in group 1 were compared with the 20 weights in group 2, because the individual kidney weights are not independent observations. A valid method would be to carry out an analysis comparing the 10 average kidney weights in group 1 with the 10 average kidney weights in group 2.

II.5.2 *Types of response variable*

Responses measured in toxicological studies can normally be classified as being one of three types:

(a) Presence/absence: A response either occurs or it does not.

(b) Ranked: A response may be present in various degrees. Thus, severity may be classed as minimal, slight, moderate, severe, or very severe.

(c) Continuous: A response may take any value, at least within a given range.

Each type of response demands a different sort of statistical technique. While analysis of presence/absence data, often referred to as "contingency table analysis", can be applied to ranked or continuous data by defining values above a given cut-off point as "present", this is not generally recommended, because it wastes information.

Continuous data are usually analysed by "parametric" methods, which assume that the statistical distribution underlying the response variable (or some transformation of it, e.g., its logarithm) has a specific form, traditionally the well-known bell-shaped Normal or Gaussian distribution. While such methods are best when the distribution assumed is correct, they can give misleading conclusions when the assumption is grossly incorrect.

For this reason, when there is doubt about the underlying distribution, it is often preferable to analyse continuous data by methods appropriate for ranked data, since these "non-parametric" methods make no such underlying assumption and their conclusions are generally more valid.

II.5.3   Types of between-group comparisons

While in the two-group study only one comparison is possible, this is not the case when more than two groups are being compared. Two particularly important types of test made in the k (> 2) group situation are the test for heterogeneity and the test for dose-related trend.

The test for heterogeneity determines whether, taken as a whole, there is significant evidence of departure from the (null) hypothesis that the groups do not differ in their effect. It is generally applicable but is not very informative, because it does not specifically take into account the likely pattern of response.

The test for dose-related trend is applicable only in studies in which the groups receive different doses of the same substance (or have some other natural ordering). It determines whether there is a tendency for response to rise in relation to the dose of the test substance. Graphically, the test for trend can be seen as determining whether a sloped straight line through the dose (x-axis)/response (y-axis) relationship fits the data significantly better than a horizontal straight line. That the trend statistic is significantly positive does not necessarily imply that the treatment increases response at all dose levels, though it is a particularly good test if the true situation is a linear non-threshold model. A trend test often detects a significant true effect when individual comparisons of treated groups with the control group fail to give significance.

Sometimes, there is significant departure from trend, e.g., when there is significant heterogeneity but no evidence of trend. This may arise because a response increases with dose at lower dose levels and then reduces at high dose levels, perhaps due to competing risks. In this situation, it may be appropriate to test for trend using only the data from the control and lower-dose groups.

II.5.4   Stratification

In the simplest situation all the animals in the study differ systematically only with respect to the treatment applied. However, often there are a number of sets of animals, each of which differs systematically only in respect to treatment, but where the characteristics of sets differ. The commonest situation relates to male and female animals, but there are many other possibilities, such as different conditions under which the response variable is measured. While it is often useful to look within each set of animals or "stratum", as it is often called, to determine the effects of treatment in each situation, it is also useful to determine whether, on the basis of the data from all the strata, an effect of treatment can be seen overall. In some situations, a relatively small number of animals in each stratum can make it difficult to pick up an effect of treatment as significant within individual strata, and a clear result can be seen only when results are combined over the strata. The essence of stratification lies in making comparisons within strata and then accumulating treatment differences over strata. Pooling data over strata and then making a single comparison can lead to erroneous conclusions. To illustrate this, consider a hypothetical study in which, in one batch of animals, 5/10 controls and 12/30 test animals responded, while in a second batch, 6/30 controls and 1/10 test animals responded. If batch were ignored, it would be noted that 11/40 controls compared with 13/40 test animals responded, leading to the erroneous conclusion that treatment tended to increase response. An appropriate analysis would consider batch as a stratifying factor and note that within both batch 1 (50% versus 40%) and batch 2 (20% versus 10%) the response in the control group was higher than that in the test group so that, combining these two differences, a conclusion would be reached that treatment tended to decrease response.

II.5.5 Age adjustment

For many conditions, such as tumour incidence, frequency increases markedly with age (and concomitantly with length of exposure to the agent), and the overall frequency in a treatment group can depend as much on the proportion of animals surviving a long time as on the actual ability of the treatment to cause the condition. To adjust for differential survival, age is usually treated as a stratifying variable so that between-group comparisons are made of animals at similar ages, results of the comparisons being combined over the different age strata. Age adjustment is normally applied to presence/absence conditions and, as discussed at length by Peto et al. (2), the correct method depends on the context of observation of the condition. There are three different situations:

(a) Conditions visible in-life: Here comparisons are made of the number of animals developing the condition in the time period as a proportion of those without the condition at the beginning of the period.

(b) Conditions visible only at death and assumed to cause death (fatal): Here comparisons are made of the number of animals dying from the condition in the time period as a proportion of those alive at the beginning of the period.

(c) Conditions visible only at death and assumed not to cause death (incidental): Here comparisons are made of the number of animals dying with the condition in the time period as a proportion of all those dying in the time period.

II.5.6 *Multiple observations per animal*

When multiple observations are made on one animal there are a number of additional types of statistical analysis, depending on the experimental situation and the objectives. It is impossible to cover all the possibilities in this summary, but the following situations are of reasonable frequency:

(a) Association between variables: Here two or more different variables are recorded on an animal and the objective is to determine whether the values are independent or are correlated.

(b) Variation in association between variables by treatment: For each treatment group an indicator of association between variables is calculated and the statistical problem is to test whether this indicator varies significantly by treatment group.

(c) Multiple observations of the same variable: For body weight and clinical chemistry data it is common to take measurements on the same animals at regular intervals throughout a long-term study. While between-group comparisons can be carried out on the basis of data at any specific point in time, this limits the amount of information in any one analysis. Statistical techniques are also available to compare groups with respect to change in response between two points in time or, more generally, with respect to the general pattern of response over a period of time.

(d) Within-animal comparisons: In most toxicological studies, different animals receive the different treatments and comparisons are made on a between-animal basis. In some studies, the same animal receives more than one treatment. In such studies, it is important to use appropriate statistical methods based on within-animal comparisons.

II.5.7 Hypothesis testing and probability values

Reports of toxicity studies often include statements such as "the relationship between treatment and blood glucose levels was statistically significant ($\underline{P}$ = 0.02)". What does this actually mean? Three points must be made.

First, there is a difference in meaning between biological and statistical significance. It is quite possible to have a relationship that is unlikely to have happened by chance and therefore statistically significant but of no biological consequence at all, the animals' well-being being unaffected. On the other hand an observation may be biologically, but not statistically, significant, such as when one or two tumours of an extremely rare type are seen in treated animals. Overall judgement of the evidence must take into account both biological and statistical significance.

Second, "$\underline{P}$ = 0.02" does not mean that the probability that there is no treatment effect is 0.02. The true meaning is that given the treatment actually had no effect whatsoever (or to phrase it more technically, under the null hypothesis) the probability of observing a difference as great or greater than that actually seen is 0.02.

Third, there are two types of probability ($\underline{P}$)-value. A "one-sided" (or one-tailed) $\underline{P}$-value is the probability of getting, by chance alone, a treatment effect in a specified direction as great or greater than that observed. A "two-sided" (two-tailed) $\underline{P}$-value is the probability of obtaining by chance alone, a treatment difference in either direction, positive or negative, as great or greater than that observed. Whenever a $\underline{P}$-value is quoted, it should be made clear which is being used. Normally, two-tailed $\underline{P}$-values are appropriate. However, when there is prior reason to expect a treatment effect in one direction only, a one-tailed $\underline{P}$-value is normally used. If a one-tailed $\underline{P}$-value is used, differences in the opposite direction to that assumed should be ignored.

While a $\underline{P}$-value of 0.001 or less can, on its own, provide very convincing evidence of a true treatment effect, less extreme $\underline{P}$-values such as $\underline{P}$ = 0.05 should be viewed as providing indicative evidence of a possible treatment effect, to be reinforced or supported by other evidence. If the difference is similar to one found in a previous study, or if the response, based on biochemical considerations, is expected, a less extreme $\underline{P}$-value would suffice than if the response was unexpected or not found at other dose levels.

Some laboratories, when presenting results of statistical analyses, assign an almost magical relevance to the 95% confidence level ($\underline{P}$ > 0.05), simply marking results significant or not significant at this level. This is very poor practice, because it gives insufficient information and does not enable the distinction between an undeniable effect and one that requires other confirmatory evidence. While it is not necessary to give P-values exactly, it is essential to give some idea of the degree of confidence. A useful method is to use plus signs to indicate positive differences (and minus signs to indicate negative differences), with ++++ meaning $P < 0.001$, +++ meaning $0.001 \leq P < 0.01$, ++ meaning $0.01 \leq P < 0.05$, and + meaning $0.05 \leq P < 0.1$. This makes it easier to assimilate findings when results for many variables are presented.

II.5.8 Multiple comparisons

Toxicological studies frequently involve making treatment/control comparisons for large numbers of variables. If no true treatment effect exists, it is possible that, purely by chance, one or more variables will show differences significant at the 95% confidence level. For example, with 100 independent variables, at least one variable would show significance 99.4% of the time. Because of this, it has been suggested that the critical value required to achieve significance should be made more stringent with increasing numbers of variables studied, so that, in testing at the 95% confidence level, 19 times out of 20, all the variables in the test show non-significance. This approach is not recommended, because frequently in toxicological studies a compound has only one or two real effects and has no effect on a large number of other variables studied. Such multiple comparison tests would make it much more difficult to demonstrate statistical significance for the real effects. In any case, there is something unsatisfactory about a situation in which the relationship between a treatment and a particular response arbitrarily depends on which other response happens to be investigated at the same time. For this reason, no reference is made below to any such procedures.

II.6 Statistical Analysis - Some Recommended Methods

A number of recommended methods of statistical analysis are listed below. For mathematical details the reader is referred to Peto et al. (2) and to Breslow & Day (3) for analysis of presence/absence data, to Siegel (4) and to Conover (5) for non-parametric analysis, and to Johnson & Leone (6) and Bennett & Franklin (7) for analysis of continuous data. When details are not found in these volumes, specific references are given.

II.6.1 Presence/absence data

II.6.1.1 Between-animal comparisons

Individual group comparisons	Fisher exact test (unstratified data) 2 x 2 corrected chi-squared test (stratified or unstratified data)
Heterogeneity	2 x k chi-squared test (stratified or unstratified data)
Dose-related trend	Armitage test (stratified or unstratified data).

See reference 3 for details of stratified tests and for tests of constancy of association over strata. See reference 2 for age-adjusted tests.

II.6.1.2 Within-animal comparisons

Individual group comparisons	McNemar test or sign test
Heterogeneity	Cochran test
Association between variables	Fisher exact test 2 x 2 corrected chi-squared test

II.6.2 Ranked data

II.6.2.1 Between-animal comparisons

Individual group comparisons	Mann Whitney U test
Heterogeneity	Kruskal-Wallis one-way analysis of variance
Dose-related trend	See reference 8

II.6.2.2 Within-animal comparisons

Individual group comparisons	Wilcoxon matched pairs signed-rank test
Heterogeneity	Friedman two-way analysis of variance
Dose-related trend	Page test

| Association between variables | Spearman's rank correlation coefficients |

II.6.3   Continuous data

Methods assume normality and homogeneity of variance between groups.

Before using the methods:

| Test for outliers | See reference 9 |
| Test for homogeneity of variance | Bartlett test |

Consider transformation of data by logarithms and/or square roots if untransformed data show heterogeneity of variance.

If variance is still heterogeneous after transformation, use methods of ranked data.

II.6.3.1   Between-animal comparisons

Individual group comparisons	Students t-test
Heterogeneity	One-way analysis of variance
Dose-related trend	Linear regression analysis

II.6.3.2   Within-animal comparisons

Individual group comparisons	Paired t-test
Heterogeneity	Two-way analysis of variance
Dose-related trend	Linear regression analysis

II.6.3.3   Association between variables

| Variation in association between variables over groups | Pearson correlation coefficient
Analysis of covariance |

| Change in variable over time | Analysis of variance to assess difference at second time-point after adjusting for first |

REFERENCES TO ANNEX II

1. FERON, V.C., GRICE, H.C., GRIESEMER, R., & PETO, R. (1980) Basic requirements for long-term assays for carcinogenicity. In: *Long-term and short-term screening assays for carcinogens: a critical appraisal*, Lyons, International Agency for Research on Cancer, pp. 21-83 (IARC Monographs on the Evaluation of the Carcinogenic Risk of Chemicals to Humans, Suppl. 2).

2. PETO, R., PIKE, M.C., DAY, N.E., GRAY, R.C., LEE, P.N., PARISH, S., PETO, J., RICHARDS, S., & WAHRENDORF, J. (1980) Guidelines for simple, sensitive significance tests for carcinogenic effects in long-term animal experiments. In: *Long-term and short-term screening assays for carcinogens: a critical appraisal*, Lyons, International Agency for Research on Cancer, pp. 311-426 (IARC Monographs on the Evaluation of the Carcinogenic Risk of Chemicals to Humans, Suppl. 2).

3. BRESLOW, N.E. & DAY, N.E. (1980) *Statistical methods in cancer research. I. The analysis of case-control studies*, Lyons, International Agency for Research on Cancer (IARC Scientific Publication No. 32).

4. SIEGEL, S. (1956) *Non-parametric statistics for the behavioural sciences*, New York, McGraw-Hill Book Company.

5. CONOVER, W.J. (1980) *Practical non-parametric statistics*, 2nd ed., New York, John Wiley and Sons.

6. JOHNSON, N.L. & LEONE, F.C. (1964) *Statistics and experimental design in engineering and the physical sciences*, New York, John Wiley and Sons, Vol. 1.

7. BENNETT, C.A. & FRANKLIN, N.L. (1954) *Statistical analysis in chemistry and the chemical industry*, New York, John Wiley and Sons.

8. MARASCUILO, L.A. & MCSWEENEY, M. (1967) Non-parametric *post hoc* comparisons for trend. *Psychol. Bull.*, 67: 401-412.

9. BARNETT, V. & LEWIS, T. (1978) *Outliers in statistical data*, New York, John Wiley and Sons.

ANNEX III. GUIDELINES FOR THE EVALUATION OF VARIOUS GROUPS OF FOOD ADDITIVES AND CONTAMINANTS

These guidelines have been established by JECFA and are reproduced here for easy reference. They are valid within the context in which they were generated, and are intended to serve as examples of guidance by JECFA for evaluating specific categories of substances.

1. **Enzyme Preparations Used in Food Processing** (adapted from: 1, p. 49; 2)

    (a) Toxicological evaluation

    For the purpose of toxicological evaluation, enzyme preparations used in food processing can be grouped into 5 major classes:

    (i) Enzymes obtained from edible tissues of animals commonly used as foods. These are regarded as foods and, consequently, considered acceptable, provided that satisfactory chemical and microbiological specifications can be established.

    (ii) Enzymes obtained from edible portions of plants. These are also regarded as foods and, consequently, considered acceptable, provided that satisfactory chemical and microbiological specifications can be established.

    (iii) Enzymes derived from microorganisms that are traditionally accepted as constituents of foods or are normally used in the preparation of foods. These products are regarded as foods and, consequently, considered acceptable, provided that satisfactory chemical and microbiological specifications can be established.

    (iv) Enzymes derived from non-pathogenic microorganisms commonly found as contaminants of foods. These materials are not considered as foods. It is necessary to establish chemical and microbiological specifications and to conduct short-term toxicity studies to ensure the absence of toxicity. Each preparation must be evaluated individually and an ADI must be established.

    (v) Enzymes derived from microorganisms that are less well known. These materials also require chemical and microbiological specifications and more extensive toxicological studies, including a long-term study in a rodent species.

Safety assessments for enzymes belonging to classes i - iii will be the same regardless of whether the enzyme is added directly to food or is used in an immobilized form. Separate situations should be considered with respect to the enzymes described in classes iv and v:

(i) Enzyme preparations added directly to food but not removed.

(ii) Enzyme preparations added to food but removed from the final product according to good manufacturing practice.

(iii) Immobilized enzyme preparations that are in contact with food only during processing.

For (i) above, an ADI should be established to ensure that levels of the enzyme product present in food are safe. The studies indicated in these guidelines are appropriate for establishing ADIs (the guidelines were originally drafted for this situation). For (ii), an ADI "not specified" may be established, provided that a large margin of safety exists between possible residues and their acceptable intake. For (iii), it may not be necessary to set an ADI for residues that could occur in food as a result of using the immobilized form of the enzyme. It is acceptable to perform the toxicity studies relating to the safety of the enzyme on the immobilized enzyme preparation, provided that information is given on the enzyme content in the preparation.

(b) Specifications for identity and purity

Prior to revising existing specifications and developing new specifications for enzyme preparations for food processing, the following data are necessary:

(i) A comprehensive description of the main enzymatic activity (or activities), including the Enzyme Commission number(s) if any.

(ii) A list of the subsidiary enzymatic activities, whether they perform a useful function or not.

(iii) A clear description of the source.

(iv) A list of non-enzymatic substances derived from the source material(s), with limits where appropriate.

(v) A list of added co-factors, with limits where appropriate.

(vi) A list of carriers and diluents, with limits where appropriate.

(vii) A list of preservatives present from manufacture or deliberately added, with limits where appropriate.

(c) <u>Immobilizing agents</u>

A number of procedures involving different chemical substances are used for immobilizing enzymes. These processes include microencapsulation (e.g., entrapment in gelatin to form an immobilized complex), immobilization by direct addition of glutaraldehyde, immobilization by entrapment in porous ceramic carrier, and complexation with agents such as DEAE-cellulose or polyethylenimine. Several agents may be used in the immobilizing process. Substances derived from the immobilizing material may be in the final product due to either the physical breakdown of the immobilized enzyme system or to impurities contained in the system. The amount of data necessary to establish the safety of the immobilizing agent depends on its chemical nature. The levels of residues in the final product are expected to be extremely low.

Some of the substances used in the preparation of immobilizing systems are extremely toxic. The levels of these substances or their contaminants permitted in the final product should be at the lowest levels that are technologically feasible, provided that these levels are below those of any toxicological concern. An ADI will not be established.

2. **Natural and Synthetic Food Colours** (Adapted from: Reference 1, Annex 6, p. 50)

For toxicological evaluation, natural colours should be considered as falling within three main groups:

(a) A colour isolated in a chemically unmodified form from a recognized foodstuff and used in the foodstuff from which it is extracted at levels normally found in that food. This product could be accepted in the same manner as the food itself with no requirement for toxicological data.

(b) A colour isolated in a chemically unmodified form from a recognized foodstuff but used at levels in excess of those normally found in that food or used in foods other than that from which it is extracted. This product might require the toxicological data usually demanded for assessing the toxicity of synthetic colours.

(c) A colour isolated from a food source and chemically modified during its production or a natural colour isolated from a non-food source. These products would also require a toxicological evaluation similar to that carried out for a synthetic colour.

It is recognized that natural colours may be reproduced by chemical synthesis but it is noted that "nature-identical" colours produced by chemical synthesis may contain impurities warranting toxicological evaluation similar to that required for a synthetically produced food colour.

The toxicological evaluation of synthetic food colours would require the following minimum data:

(a) Metabolic studies in several species, preferably including man. These should include studies on absorption, distribution, biotransformation, and elimination, and an attempt should be made to identify the metabolic products in each of these steps.

(b) Short-term feeding studies in a non-rodent mammalian species.

(c) Multi-generation reproduction/teratogenicity studies.

(d) Long-term carcinogenicity/toxicity studies in two species.

3. **Solvents Used in Food Processing** (Adapted from: Reference 1, Annex 6, pp. 50-51)

Extraction solvents are used *inter alia* in the extraction of fats and oils, defatting fish and other meals, and in decaffeinating coffee and tea. They are chosen mainly for their ability to dissolve the desired food constituents selectively and for their volatility, which enables them to separate easily from the extracted material with minimum damage. The points raised by their use relate to:

(a) toxicity of their residues;

(b) toxicity of any impurities in them;

(c) toxicity of substances such as solvent stabilizers and impurities that may be left behind after the solvent is removed; and

(d) toxicity of any substances produced as a result of a reaction between the solvent and food ingredients.

Before any extraction solvent can be evaluated, information is required on:

(a) identity and amount of impurities in the solvent (including those that are formed, acquired, or concentrated owing to continuous reuse of the solvent);

(b) identity and amount of stabilizers and other additives; and

(c) toxicity of residues of solvents, additives, and impurities.

Impurities are particularly important because there are wide differences in the purities of food grade and industrial grade solvents. The food use of extraction solvents is frequently much less than the industrial use, and hence their food-grade requirements may receive insufficient consideration, both in food use and in toxicological testing. Furthermore, the impurities or stabilizers may not have the same volatility as the solvent itself, and as a result, these may be left behind in the food after the solvent is removed. Finally, the possibility of any solvent, impurity, stabilizer, or additive reacting with food ingredients should be checked.

When biological and toxicological data raise doubts about a substance's safety, two approaches are possible: (a) to set an ADI for the substance or (b) to discourage its use altogether. Even when data indicate a wide margin of safety for a substance, or when there is a paucity of toxicological data on the substance, but no problems concerning the impurities, residues, and any chemical reaction with food ingredients, it would be appropriate to limit the use of the substance to the minimum possible level.

When the data on a substance indicate the presence of certain impurities in the tested material, considerable problems arise in its evaluation. This is especially true if industrial-grade rather than food-grade material has been used in the toxicological study. For example, when evaluating the solvents 1,1,1-trichloroethane, trichloroethylene, and tetrachloroethylene, it was noted that the toxicological data indicated the presence of certain known toxic and carcinogenic substances. The interpretation of these data became extremely difficult because industrial-grade material had been used in the studies. Only food-grade material should be used in toxicological studies and the impurities in the material should be fully identified.

Carrier solvents raise somewhat different issues. They are used for dissolving and dispersing nutrients, flavours, antioxidants, emulsifiers, and a wide variety of other food ingredients and additives. With the exception of carrier solvents for flavours, they tend to occur at higher levels in food than extraction solvents, mainly because frequently no attempt is made to remove them, and because some of them are relatively nonvolatile. Since carrier solvents are intentional additives and are often not removed from the processed food, it is important to evaluate their safety together with the safety of any additives or stabilizers in them.

4. **Residues Arising from Use of Xenobiotic Anabolic Agents in Animal Feed** (Adapted from: Reference 1, p. 13 and reference 3, p. 15)

Many studies have established the importance and efficacy of anabolic agents for meat production. Two categories of compounds are used - namely:

(a) hormones that are identical to those occurring naturally in food-producing animals and human beings, including the esters of these hormones; and

(b) xenobiotic compounds, such as derivatives of hormones, synthetic compounds with hormonal activity, natural-product hormonally active agents that are not identical with human endogenous hormones, and derivatives of such compounds.

The toxicological evaluation of residues of anabolic agents that are present in human food obtained from animals treated with these agents must take into account whether the residue is identical to a human endocrine hormone. In the latter case, the possible endocrinological effects and carcinogenic potential of the residue must be closely examined.

Chemically modified hormones, hormonally active agents from plants, and synthetic anabolic agents present the following specific problems:

(a) extreme potency and consequently the need to ensure minimal residues;

(b) potential tumorigenic activity; and

(c) the presence of their metabolites in animal products that might be of endocrinological or toxicological consequence.

The evaluation for acceptance of the use of xenobiotic anabolic agents in animal food production resembles in many respects the evaluation of pesticides, since the two essential elements required are:

(a) adequate, relevant toxicological data, and

(b) comprehensive data about the kinds and levels of residues when the substances are used in accordance with good animal husbandry practice, which requires evidence as to the efficacy of the anabolic agent, the amounts used to produce the effect, the residue levels based on field trials, and information about methods of analysis of residue levels that could be used for control or monitoring purposes.

5. **Metals in Food** (Adapted from: Reference 1, pp. 14-15)

Toxicological evaluation of metals in foods calls for carefully balanced consideration of the following factors:

(a) nutritional requirements, including nutritional interactions with other constituents of food (including other metals when the interactions are nutritionally or toxicologically relevant) in respect of, for instance, absorption, storage in the body, and elimination;

(b) the results of epidemiological surveys and formal toxicological studies, including interactions with other constituents of food (including other metals when the interactions are nutritionally or toxicologically relevant), information about pharmaceutical and other medicinal uses, and clinical observations on acute and chronic toxicity in human experience and veterinary practice;

(c) total intake on an appropriate time basis (e.g., daily, weekly, yearly or lifetime) from all sources (food, water, air) of metals as normal constituents of the environment, as environmental contaminants, and as food additives of an adventitious or deliberate nature.

The tentative tolerable daily intakes proposed for certain metals by the Committee provide a guideline for maximum tolerable exposure. In the case of essential elements, these levels exceed the normal daily requirements, but this should not be construed as an indication of any change in the recommended daily requirements. In the case of both essential and non-essential metals, the tentative tolerable intake reflects permissible human exposures to these substances as a result of natural occurrence in foods or various food processing practices, as well as exposure from drinking-water.

It is important that the proposed tolerable intakes are not used as guidelines for fortifying processed food, since the currently accepted values for required daily intake are sufficient to meet the known nutritional requirements.

6. Packaging Materials (Adapted from: Reference 4, pp. 22-23)

Many substances may become food contaminants as a result of the use of food-contact materials. The Committee considered that the general principles governing the use of food additives established by the WHO Scientific Group on Procedures for Investigating Intentional and Unintentional Food Additives, and the WHO Scientific Group on the Assessment of the Carcinogenicity and Mutagenicity of Chemicals, should be applied in the overall evaluation of substances migrating into food from food-contact materials.

Many such materials are made of polymer systems and the polymers themselves are usually inert, non-toxic, and do not migrate into food. However, monomers, which are inevitably present in the polymeric materials, residual reactants, intermediates, manufacturing aids, solvents, and plastic additives, as well as the products of side reactions and chemical degradation, may be present. These substances may migrate into food and may be toxic. Migration from food-contact materials may arise during processing, storage, and preparation operations such as heating, microwave cooking, or treatment with ionizing radiation. For evaluation purposes, information on the following is required:

(a) the chemical identity and toxicological status of the substances that enter food;

(b) the possible exposure, details of which can be derived from migration studies using suitable extraction procedures, and/or the analysis of food samples; and

(c) the nature and amount of food in contact with the packaging materials, and the intake of such food.

It is necessary, in many instances, to recommend that human exposure to migrants from food-contact materials be restricted to the lowest levels technologically attainable. One way to achieve this is to draw up strict specifications limiting the quantities in the materials. It is also necessary to determine whether food processing has an effect in generating the potentially toxic substances in food-contact materials.

Generally, evaluation will require the following:

(a) the lowest levels of potential migrants from within the polymeric system(s) that are technologically attainable with improved manufacturing processes for food-contact materials;

(b) the resulting levels of the migrants in food;

(c) the intake of the foods; and

(d) the most appropriate statistical design that will enable the implications for health to be interpreted from adequate and relevant toxicological data.

A monitoring programme should be established, with a view to supplementing the existing data on human exposure and providing a means of demonstrating a reduction in such exposure as techniques improve. Priority in the programme should be given to the substances with the greatest potential for adversely affecting human health.

REFERENCES TO ANNEX III

1. FAO/WHO (1982) Evaluation of certain food additives and contaminants. Twenty-sixth report of the Joint FAO/WHO Expert Committee on Food Additives (WHO Technical Report Series No. 683).

2. FAO/WHO (1986) Evaluation of certain food additives and contaminants. Twenty-ninth report of the Joint FAO/WHO Expert Committee on Food Additives (WHO Technical Report Series No. 733).

3. FAO/WHO (1981) Evaluation of certain food additives. Twenty-fifth report of the Joint FAO/WHO Expert Committee on Food Additives (WHO Technical Report Series No. 669).

4. FAO/WHO (1984) Evaluation of certain food additives and contaminants. Twenty-eighth report of the Joint FAO/WHO Expert Committee on Food Additives (WHO Technical Report Series No. 710).

ANNEX IV. EXAMPLES OF THE USE OF METABOLIC STUDIES IN THE SAFETY ASSESSMENT OF FOOD ADDITIVES

As indicated in section 5.2, it is not feasible or desirable to develop simple guidelines for pharmacokinetic and metabolic studies. In view of this, the examples given below of several food additives and contaminants on which a great deal of biochemical work has been done will serve to highlight the value and many of the problems associated with the use of these studies for determining mechanisms. Clearly, further research will be needed to solve these problems.

1. Sodium Cyclamate

This compound represents a unique situation in toxicology in that it has been generally agreed that levels of a metabolite rather than the parent compound should be used for the usual safety determinations. The twenty-sixth Committee allocated an ADI of 0 - 11 mg/kg body weight to cyclamate, calcium and sodium salts, expressed as cyclamic acid (1).

It has been shown that a metabolite of cyclamate, cyclohexylamine (an active pressor amine), appears in the urine in variable amounts after variable time intervals from the administration of cyclamate in rats (2, 3). This metabolite has been found to be produced by bacterial action in the intestine (3, 4), but only after intestinal flora have undergone undefined adaptive changes (2). Therefore, it normally appears only after a latent period. However, in certain human subjects, some immediate converters have been found to be present (2, 5). In both animals and man, the ability of intestinal flora to convert cyclamate to cyclohexylamine varies widely with time in the same individual. The number of individuals able to convert cyclamate to cyclohexylamine and the level at which this conversion occurs have been factored into ADI determinations using averages from some studies (6). However, it is difficult to obtain really consistent figures, and those in use represent compromises. The ADI is based on subsequent studies, which demonstrated that cyclohexylamine, administered orally, induced testicular atrophy in rats (7, 8).

The primary reason for the decision by JECFA and various national regulatory bodies to agree to the use of the levels of this metabolite rather than the parent compound for toxicological evaluation purposes appears to have been the nature of the metabolite, in this instance, a compound that is pharmacologically-active relative to the parent compound and is capable of inducing testicular atrophy in rats (9). However, it is questionable that the readiness of all these bodies to accept this unique approach in this situation is, in fact, appropriate.

The presence of the metabolite certainly cannot be ignored; however, it would seem more logical to demand that the effects in question (testicular atrophy) be demonstrated following feeding the parent compound (cyclamate). This approach is complicated by the inconsistent nature of the appearance of the metabolite.

The individuality of the response has necessitated a most conservative attitude that has raised the important general question of how the toxicologist can best protect vulnerable individuals. In addition, this example has pointed to the importance of studying metabolism by gut flora in toxicological evaluations. Unquestionably, variations in gut flora are one of the more important determinants of species differences, and the example of cyclamate has pointed the way to studies in which this factor has been reduced to a minimum.

2. Sodium Saccharin

JECFA has reviewed the safety of saccharin many times. In 1984, the twenty-eighth Committee allocated a temporary group ADI of 0 - 2.5 mg/kg body weight to saccharin, including its calcium, potassium, and sodium salts (10). Subsequent to the demonstration in some long-term toxicity studies that sodium saccharin could induce tumours of the urinary bladder in male rats at high dose levels, much work was undertaken in an effort to explain this phenomenon. Only two of the many reported findings will be discussed in this section, as illustrations of certain general principles.

On the basis of the first series of studies, it would seem unquestionable that sodium saccharin is not metabolized in the rat; this seems to be generally applicable to human beings and other species (11). There is no postulated theoretical mechanism of chemical carcinogenesis that could fit this picture. The second series of studies have established the most interesting finding that, although saccharin is not metabolized, it can modify the metabolic pathway of certain normal constituents in the diet. A dose-related increase in certain tryptophan metabolites - notably indoxyl sulfate - was found in the urine of saccharin-treated rats (12). In contrast, these effects could not be demonstrated in human beings consuming saccharin (13). In view of previous interest in the association of tryptophan and its metabolites with bladder tumour induction and its occurrence in man, this observation was of great interest. Although the further steps in this series of investigations have not succeeded in establishing a convincing relationship between the carcinogenicity finding and the metabolism of tryptophan, nevertheless, a most important general principle in toxicology has been demonstrated that remains to be exploited. The fact that a compound that is not metabolized could change the metabolism of other compounds provides a basis for studies of mechanism of action not considered in the past. It seemed likely, initially, that the possible bacteriostatic action of saccharin might be affecting the gut flora; although this seems to be a possible practical explanation, further study is needed to explain the whole picture.

3. o-Phenylphenol (OPP)

This compound is a fungicide widely used on oranges, the use of which results in low residue levels as a food contaminant. The Joint Meeting on Pesticide Residues (JMPR) has allocated a temporary ADI to OPP (and its sodium salt) of 0 - 0.02 mg/kg body weight (14). It has been chosen as an example of a situation in which extensive metabolic studies have been correlated with toxicological findings.

OPP has been found to give rise to tumours of the urinary bladder when fed to rats at relatively high levels in the diet (15). Two metabolites, the glucuronide conjugate of OPP and the sulfate ester conjugate of OPP, have been identified in the urine of rats after administration of different levels of OPP. A third metabolite, tentatively postulated to be a conjugated dihydroxybiphenyl compound, has been reported after the administration of a high level of OPP, the same level required to induce bladder tumours, but not after the administration of lower levels of OPP (16). From this observation, some investigators have concluded that, at lower levels of administration, carcinogenicity does not occur, because the "proximate carcinogen" (i.e., the high-level metabolite) has not been formed. This fascinating study is, unfortunately, incomplete. There are disputes as to the absence of the "active" metabolite (the conjugated dihydroxybiphenyl compound) at the lower dose levels; the detection limits of the various metabolites, which are not clearly delineated in the available literature, are a matter of considerable concern. In addition, the "active" metabolite has no special chemical characteristics to suggest that it would conform to any of the current theories of action of chemical carcinogens. These results show the difficulties of proving the mechanism of a tumourigen so that a safe dose can be established, even though the available evidence points in that direction. It is clear that much more information is needed before the postulated change in metabolism can be related to carcinogenesis.

4. Methylene Chloride

This compound, which is used as a food extraction solvent in some countries, has been the subject of intensive carcinogenicity, mutagenicity, metabolism, pharmacokinetic, and epidemiological studies, but questions still exist about its safety of use. Because of the inadequacies of the studies available at that time, the twenty-sixth Committee withdrew the previously-allocated ADI for methylene chloride (17).

The safety of methylene chloride was brought into question by a long-term inhalation study that produced very rare salivary gland sarcomas in male rats in an apparent dose-response relationship at 1500 and 3500 mg/kg air; an increased incidence of tumours was not observed in female rats or in hamsters in a parallel study (unpublished studies by the Dow Chemical Company, Midland, Michigan, USA, 1980). Also, preliminary results from a mouse inhalation study indicate an increased incidence of liver and lung neoplasms at 1000 and 4000 mg methylene chloride/kg air (18). However, arguments have been made on the basis of metabolic studies that the maximum tolerated dose (MTD) was exceeded in these long-term inhalation studies in rats and mice (19). Bioassays in which methylene chloride was administered at lower levels by inhalation or in drinking-water have not resulted in a significant increase in malignant tumours (unpublished studies by the Dow Chemical Company, Midland, Michigan, USA, 1982 and by the National Coffee Association, USA, 1982).

Methylene chloride is metabolized via two pathways. The principle site of metabolism is the liver in all species studied, including man. One pathway involves glutathione, giving rise to formaldehyde, which is oxidized to formic acid and then carbon dioxide. The other pathway is mediated by cytochrome P-450 and involves dehydrochlorination to carbon monoxide and hydrogen chloride. One of the intermediates in the first pathway, a glutathione conjugate, has been implicated as the DNA-reactive species responsible for the apparent mutagenicity of methylene chloride in some tests. However, there is no evidence of alkylation in animals (20, 21, 22).

There is a linear discontinuity in metabolite formation (carbon dioxide and carbon monoxide) as exposure to methylene chloride increases. For example, it has been shown that, on inhalation of 174 mg methylene chloride/m^3 (50 ppm), 95% is metabolized, while at 1750 mg/m^3 air (500 ppm), only 69% is metabolized and at 5200 mg methylene chloride/m^3 air (1500 ppm), only 45% is metabolized to carbon dioxide and carbon monoxide. Both oral and inhalation studies show that saturation of metabolism occurs in all species examined (rat, mouse, hamster, and man) (19).

Greater amounts of methylene chloride are metabolized when the compound is presented in drinking water than when the same daily dose is gavaged in a single dose, either in corn oil or in water. Administration of a large number of doses in small amounts, such as when methylene chloride is administered in the drinking-water, yields greater amounts of metabolites than when the total amount is given at one time, such as by gavage. The vehicle used in gavage studies also plays a role in the clearance of methylene chloride from various tissues; for example, the compound is dissipated from both blood and liver in less than 2 h after administration by gavage in water compared with a residence time of about 8 h in venous blood and over 25 h in the liver after administration by gavage in corn oil (19).

Despite tremendous efforts to study this compound biochemically, a clear picture of the mechanism of its biological effects has not emerged. This shows the difficulties of developing sufficient biochemical data to set a safe dose level for a substance that causes cancer at high dose levels. The saturation effect and the occurrence of tumours at high dose levels may be related. An encouraging point is that primates metabolize chlorinated solvents to a lesser extent than rats or mice; thus, less of the glutathione reactive intermediate, which has been postulated as being responsible for the genotoxic effects of methylene chloride, should be present in man than in the animals exhibiting the deleterious effects. Finally, the differences in rates of metabolism of methylene chloride, depending on the route of administration, point to the need for very careful assessment of the appropriate route of administration to mimic exposure in man.

5. Trichloroethylene

This chemical is an extraction solvent. It has been reviewed by JECFA, but an ADI has not been allocated (17). Trichloroethylene has been found to cause an increased incidence of hepatocellular carcinomas in mice, but not in rats (23,24). The earlier bioassays were performed with industrial-grade trichloroethylene, which contained epoxide stabilizers, at least one of which is a potent mutagen (section 3 in Annex III). However, the results of later studies using non-epoxide stabilized material confirmed the results of the earlier studies, indicating that the stabilizers were not responsible for the hepatocarcinogenicity observed in mice.

Trichloroethylene has been subjected to a great deal of metabolic and pharmacokinetic research on mice and rats, and an interesting story is emerging, which could explain the difference in response between these species and the relevance to man of the rodent bioassays.

Electron micrographs of liver tissue from mice that had been dosed with a high level of trichloroethylene for 10 days showed a proliferation of peroxisomes; significant proliferation was not observed in rats after the administration of the same amount of trichloroethylene (25). Biochemical studies have shown that cyanide-insensitive acyl CoA oxidase, an enzyme present within the peroxisome that ultimately produces hydrogen peroxide, is enhanced 5 - 16-fold in the peroxisome-proliferated state compared within the control. However, only a small increase in catalase activity, which catalyses the conversion of hydrogen peroxide to water, has been observed in the proliferated cell (25, 26, 27). It has been postulated that the large increase in acyl CoA oxidase activity coupled with the marginal increase in catalase activity leads to an increased steady-state concentration of hydrogen peroxide within the liver cell, which results in cytotoxicity and DNA damage and eventually cancer in mice (28).

Trichloroethylene appears to be metabolized via the cytochrome P-450 system, in both the rat and mouse (29, 30). The major metabolite, which is ultimately converted to carbon dioxide, is trichloroacetic acid. Apparently, the enzymes in this pathway are not induced to the same extent in rats as in mice, as shown by gavage studies with trichloroethylene; mice show linear kinetics with respect to metabolite formation over a wide dose range, while rats show saturation kinetics; saturation is observed at low levels relative to amounts required to observed peroxisome proliferation and the induction of hepatocellular carcinomas in mice. In contrast, when trichloroacetic acid is administered to rats and mice, a similar dose-dependent large increase in peroxisomes is observed in both species. This suggests that trichloroacetic acid is the proximate peroxisome proliferator, and the reason that proliferation is not observed after the administration of trichloroethylene to rats is that not enough of the acid is produced to cause the effect. If peroxisome proliferation (and a consequent increased steady-state concentration of hydrogen peroxide in the cell) is responsible for hepatocellular carcinoma induction, then trichloroacetic acid should be a hepatocellular carcinogen in both species. This hypothesis is at presently being tested (28).

Where does man fit into the picture? Preliminary work with hepatocytes isolated from mice, rats, and man show that the human hepatocyte is much more like that of the rat than that of the mouse in terms of its ability to convert trichloroethylene to trichloroacetic acid; in fact, the human hepatocyte is even less active in converting trichloroethylene to trichloroacetic acid than that of the rat (28). In addition, when trichloroacetic acid is administered to cultured human hepatocytes, there is no evidence of peroxisome proliferation, as measured by cyanide-insensitive acyl CoA oxidase activity. In contrast, studies on mouse and rat hepatocytes have registered large, dose-related increases in this enzyme activity. These data suggest that the mouse bioassay data showing an increase in hepatocellular carcinomas after the administration of trichloroethylene are not appropriate for human beings because:

(a) trichloroethylene is not converted into trichloroacetic acid at a high enough rate in man to cause peroxisome proliferation; and

(b) trichloroacetic acid does not appear to cause peroxisome proliferation in man at the levels at which it causes the effect in rats and mice.

Further data are needed to obtain a firm conclusion that could withstand regulatory scrutiny.

6. Estragole

This chemical is a naturally-occurring anisole derivative that is used as a flavouring agent. Estragole has been reviewed by JECFA, but an ADI has not been allocated (31). It has been found to be a mouse carcinogen at a dose level of approximately 500 mg/kg body weight per day (32, 33). In contrast, the estimated human daily intake of estragole in the diet is approximately 1 µg/kg body weight (349).

One of the routes of metabolism of estragole is through a hydroxylated intermediate, 1'-hydroxyestragole (35). This "activated" intermediate is likely to undergo esterification reactions with cellular constituents to form electrophilic conjugates. It has been postulated that the carcinogenic effect seen with estragole is due to the formation of the "proximate carcinogen", 1'-hydroxyestragole, which reacts to form the "ultimate carcinogen", the electrophilic conjugate (32, 33).

Studies have been performed in which the level of 1'-hydroxyestragole has been measured in the urine of mice exposed to various levels of estragole. At the low dose, 0.5 mg/kg body weight, 1 - 2% of the dose was excreted as 1'-hydroxyestragole, while, at the high dose, 1000 mg/kg body weight, 10 - 15% of the ingested dose of estragole was excreted as the hydroxylated compound. In studies on human volunteers, fed 1 µg estragole, 0.3% of the dose was excreted as 1'-hydroxyestragole (34).

These and other data suggest that only at very high and overwhelming levels of estragole are significant amounts of the activated intermediate formed, and that there appear to be over 6 orders of magnitude difference between levels of the intermediate in the high-dose mouse study and the level present in man at normal levels of consumption. This hypothesis, if it holds, leads to the conclusion that the human carcinogenic risk from the ingestion of normal levels of estragole is negligible.

REFERENCES TO ANNEX IV

1. FAO/WHO (1982) Evaluation of certain food additives and contaminants. Twenty-sixth report of the Joint FAO/WHO Expert Committee on Food Additives (WHO Technical Report Series No. 683).

2. RENWICK, A.G. & WILLIAMS, R.T. (1972) The fate of cyclamate in man and other species. Biochem. J., 129: 869-879.

3. BICKEL, M.H., BURKARD, B., MEIER-STRASSER, E., & VAN DEN BROEK-BOOT, M. (1974) Entero-bacterial formation of cyclohexylamine in rats ingesting cyclamate. Xenobiotica, 4: 425-439.

4. DRASSAR, B.S., RENWICK, A.G., & WILLIAMS, R.T. (1972) The role of the gut flora in the metabolism of cyclamate. Biochem. J., 129: 881-890.

5. ASAHINA, M., YAMAHA, T., WATANABE, K., & SARRAZIN, G. (1971) Excretion of cyclohexylamine, a metabolite of cyclamate, in human urine. Chem. Pharm. Bull. (Tokyo), 19: 628-632.

6. RENWICK, A.G. (1983) The fate of cyclamate in man and rat. In: Transcripts of the European Toxicological Forum, 18-22 October 1983, Geneva, pp. 301-312.

7. GAUNT, I.F., SHARRATT, M., GRASSO, P., LANSDOWN, A.B.G., & GANGOLLI, S.D. (1974) Short-term toxicity of cyclohexylamine hydrochloride in the rat. Food Cosmet. Toxicol., 12: 609.

8. MASON, P.L. & THOMPSON, G.R. (1977) Testicular effects of cyclohexylamine hydrochloride in the rat. Toxicology, 8: 143.

9. FAO/WHO (1978) Evaluation of certain food additives. Twenty-first report of the Joint FAO/WHO Expert Committee on Food Additives (WHO Technical Report Series No. 617).

10. FAO/WHO (1984) Evaluation of certain food additives and contaminants. Twenty-eighth report of the Joint FAO/WHO Expert Committee on Food Additives (WHO Technical Report Series No. 710).

11. RENWICK, A.G. (1985) The disposition of saccharin in animals and man: a review. Food Chem. Toxicol., 23: 429-435.

12. SIMS, J. & RENWICK, A.G. (1985) The microbial metabolism of tryptophan in rats fed a diet containing 7.5% saccharin in a two-generation protocol. Food Chem. Toxicol., 23: 437-444.

13. ROBERTS, A. & RENWICK, A.G. (1985) The effect of saccharin on the microbial metabolism of tryptophan in man. Food Chem. Toxicol., 23: 451-455.

14. FAO/WHO (1985) Pesticide residues in food. Report of the Joint Meeting of the FAO Panel of Experts on Pesticide Residues in Food and the Environment and a WHO Expert Group on Pesticide Residues (FAO Plant Production and Protection Paper No. 68).

15. HIRAGA, K. & FUJII, T. (1984) Induction of tumours of the urinary bladder in F-344 rats by dietary administration of o-phenylphenol. Food Cosmet. Toxicol., 22: 865-870.

16. REITZ, R.H., FOX, T.R., QUAST, J.F., HERMANN, E.A., & WATANABE, P.G. (1983) Molecular mechanisms involved in the toxicity of orthophenylphenol and its sodium salt. Chem.-Biol. Interact., 43: 99-119.

17. FAO/WHO (1983) Evaluation of certain food additives and contaminants. Twenty-seventh report of the Joint FAO/WHO Expert Committee on Food Additives (WHO Technical Report Series No. 696 and corrigenda).

18. FOOD CHEMICAL NEWS (1985) Methylene chloride is carcinogenic in inhalation tests conducted for NTP, March 25, p. 38.

19. KIRSCHMAN, J. (1984) Methylene chloride: safety testing overview. In: Transcripts of the European Toxicology Forum, 18-21 September 1984, Geneva, pp. 197-210.

20. AHMED, A. & ANDERS, M. (1976) Metabolism of dihalomethanes to formaldehyde and inorganic halide. I. *In vitro* studies. Drug Metab. Dispos., 4: 357-361.

21. AHMED, A., ET AL. (1980) Halogenated methanes: metabolism and toxicity. Fed. Proc., 39: 3150-3155.

22. GREEN, T. (1984) Genotoxicity of methylene chloride. In: Transcripts of the European Toxicology Forum, 18-21 September 1984, Geneva, pp. 211-215.

23. NCI (1976) Carcinogenesis bioassay of trichloroethylene (CAS Registry No. 79-01-6), National Cancer Institute (DHEW Publication No. (NIH) 76-802).

24. NTP (1983) Draft report abstracts on nine chemical carcinogenesis bioassays, National Toxicology Program, pp. 767-768 (Chemical Regulation Report No. 6).

25. ELCOMBE, C.R., ROSE, M.S., & PRATT, I.S. (1985) Biochemical, histological, and ultrastructural changes in rat and mouse liver following the administration of trichloroethylene: possible relevance to species differences in hepatocarcinogenicity. Toxicol. appl. Pharmacol., 79: 365-376.

26. LALWANI, N.D., REDDY, M.K., MANGKORNKANOL-MARK, M., & REDDY, J.K. (1981) Induction, immunochemical identity and immunofluorescence localization of an 80 000-molecular-weight peroxisome-proliferation-associated polypeptide (polypeptide PPA-80) and peroxisomal enoyl-CoA hydratase of mouse liver and renal cortex. Biochem. J., 198: 177-186.

27. REDDY, J.K., LALWANI, N.D., QURESHI, S.A., REDDY, M.K., & MOEHLE, C.M. (1984) Induction of hepatic peroxisome proliferation in non-rodent species, including primates. Am. J. Pathol., 114: 171-183.

28. ELCOMBE, C.R. (1984) In: Transcripts of the European Toxicology Forum, 18-21 September 1984, Geneva, pp. 134-144.

29. GREEN, T. & PROUT, M.S. (1985) Species differences in response to trichloroethylene. II. Biotransformation in rats and mice. Toxicol. appl. Pharmacol., 401-411.

30. PROUT, M.S., PROVAN, W.M., & GREEN, T. (1985) Species differences in response to trichloroethylene. I. Pharmacokinetics in rats and mice. Toxicol. appl. Pharmacol., 79: 389-400.

31. FAO/WHO (1981) Evaluation of certain food additives. Twenty-fifth report of the Joint FAO/WHO Expert Committee on Food Additives (WHO Technical Report Series No. 669).

32. DRINKWATER, N.R., MILLER, E.C., MILLER, J.A., & PITOT, H.C. (1976) The hepatocarcinogenicity of estragole (1-allyl-r-methoxybenzene) and 1'-hydroxyestragole in the mouse and the mutagenicity of 1'-acetoxyestragole in bacteria. J. Natl Cancer Inst., 57: 1323-1331.

33. MILLER, E.C., SWANSON, A.B., PHILLIPS, D.H., FLETCHER, T.L., LIEM, A., & MILLER, J.A. (1983) Structure-activity studies of the carcinogenicities in the mouse and rat of some naturally occurring and synthetic alkenylbenzene derivatives related to safrole and estragole. Cancer Res., 43: 1124-1134.

34. BERNARD, B. & CALDWELL, J. (1984) Significance of dose-dependent metabolism of flavour chemicals for their safety evaluation. In: Transcripts of the European Toxicology Forum, 18-21 September 1984, Geneva, pp. 124-131.

35. SWANSON, A.B., MILLER, E.C., & MILLER, J.A. (1981) The side-chain epoxidation and hydroxylation of the hepatocarcinogens safrole and estragole and some related compounds by rat and mouse liver microsomes. Biochim. Biophys. Acta, 673: 505-516.

ANNEX V. APPROXIMATE RELATION OF PARTS PER MILLION IN THE DIET TO MG/KG BODY WEIGHT PER DAY[a]

Animal	Weight (kg)	Food consumed per day (g) (liquids omitted)	Type of diet	1 ppm in food = (mg/kg body weight per day)	1 mg/kg body weight per day = (ppm of diet)
Mouse	0.02	3		0.150	7
Chick	0.40	50		0.125	8
Rat (young)	0.10	10	Dry laboratory chow diets	0.100	10
Rat (old)	0.40	20		0.050	20
Guinea-pig	0.75	30		0.040	25
Rabbit	2.0	60		0.030	33
Dog	10.0	250		0.025	40
Cat	2	100	Moist, semi-solid diets	0.050	20
Monkey	5	250		0.050	20
Dog	10	750		0.075	13
Man	60	1500		0.025	40
Pig or sheep	60	2400	Relatively dry grain forage mixtures	0.040	25
Cow (maintenance)	500	7500		0.015	65
Cow (fattening)	500	15 000		0.030	33
Horse	500	10 000		0.020	50

[a] Lehman, A.J. (1954) *Association of Food and Drug Officials Quarterly Bulletin*, 18: 66. The values in this table are average figures, derived from numerous sources.

Example: What is the value in ppm and mg/kg body weight per day of 0.5% substance X mixed in the diet of a rat?

Solution: I. 0.5% corresponds to 5000 ppm.

II. From the table, 1 ppm in the diet of a rat is equivalent to 0.050 mg/kg body weight per day. Consequently, 5000 ppm is equivalent to 250 mg/kg body weight per day (5000 x 0.050).

ANNEX VI. REPORTS AND OTHER DOCUMENTS RESULTING FROM MEETINGS OF THE JOINT FAO/WHO EXPERT COMMITTEE ON FOOD ADDITIVES

1. FAO/WHO (1957) <u>General principles governing the use of food additives. First report of the Joint FAO/WHO Expert Committee on Food Additives</u> (FAO Nutrition Meetings Report Series No. 15; WHO Technical Report Series No. 129) (out of print).

2. FAO/WHO (1958) <u>Procedures for the testing of intentional food additives to establish their safety for use. Second report of the Joint FAO/WHO Expert Committee on Food Additives</u> (FAO Nutrition Meetings Report Series No. 17; WHO Technical Report Series No. 144) (out of print).

3. FAO/WHO (1962) <u>Specifications for identity and purity of food additives (microbial preservatives and antioxidants). Third report of the Joint FAO/WHO Expert Committee on Food Additives</u>. These specifications were subsequently revised and published as <u>Specifications for identity and purity of food additives. I. Antimicrobial preservatives and antioxidants</u>, Rome, Food and Agricultural Organization of the United Nations (out of print).

4. FAO/WHO (1963) <u>Specifications for identity and purity of food additives (food colours). Fourth report of the Joint FAO/WHO Expert Committee on Food Additives</u>. These specifications were subsequently revised and published as <u>Specifications for identity and purity of food additives. II. Food colours</u>, Rome, Food and Agricultural Organization of the United Nations (out of print).

5. FAO/WHO (1961) <u>Evaluation of the carcinogenic hazards of food additives. Fifth report of the Joint FAO/WHO Expert Committee on Food Additives</u> (FAO Nutrition Meetings Report Series No. 29; WHO Technical Report Series No. 220) (out of print).

6. FAO/WHO (1962) <u>Evaluation of the toxicity of a number of antimicrobials and antioxidants. Sixth report of the Joint FAO/WHO Expert Committee on Food Additives</u> (FAO Nutrition Meetings Report Series No. 31; WHO Technical Report Series No. 228) (out of print).

7. FAO/WHO (1964) <u>Specifications for the identity and purity of food additives and their toxicological evaluation: emulsifiers, stabilizers, bleaching and maturing agents. Seventh report of the Joint FAO/WHO Expert Committee on Food</u>

Additives (FAO Nutrition Meetings Report Series No. 35; WHO Technical Report Series No. 281) (out of print).

8. FAO/WHO (1965) Specifications for the identity and purity of food additives and their toxicological evaluation: food colours and some antimicrobials and antioxidants. Eighth report of the Joint FAO/WHO Expert Committee on Food Additives (FAO Nutrition Meetings Report Series No. 38; WHO Technical Report Series No. 309) (out of print).

9. FAO/WHO (1965) Specifications for identity and purity and toxicological evaluation of some antimicrobials and antioxidants (FAO Nutrition Meetings Report Series No. 38A; WHO/Food Add/24.65 (out of print).

10. FAO/WHO (1966) Specifications for identity and purity and toxicological evaluation of food colours (FAO Nutrition Meetings Report Series No. 38B; WHO/Food Add/66.25).

11. FAO/WHO (1966) Specifications for the identity and purity of food additives and their toxicological evaluation: some antimicrobials, antioxidants, emulsifiers, stabilizers, flour-treatment agents, acids, and bases. Ninth report of the Joint FAO/WHO Expert Committee on Food Additives (FAO Nutrition Meetings Report Series No. 40; WHO Technical Report Series No. 339) (out of print).

12. FAO/WHO (1967) Toxicological evaluation of some antimicrobials, antioxidants, emulsifiers, stabilizers, flour-treatment agents, acids, and bases (FAO Nutrition Meetings Report Series No. 40A,B,C; WHO/Food Add/67.29).

13. FAO/WHO (1967) Specifications for the identity and purity of food additives and their toxicological evaluation: some emulsifiers and stabilizers and certain other substances. Tenth report of the Joint FAO/WHO Expert Committee on Food Additives (FAO Nutrition Meetings Report Series No. 43; WHO Technical Report Series No. 373).

14. FAO/WHO (1968) Specifications for the identity and purity of food additives and their toxicological evaluation: some flavouring substances and non-nutritive sweetening agents. Eleventh report of the Joint FAO/WHO Expert Committee on Food Additives (FAO Nutrition Meetings Report Series No. 44; WHO Technical Report Series No. 383).

15. FAO/WHO (1968) Toxicological evaluation of some flavouring substances and non-nutritive sweetening agents (FAO Nutrition Meetings Report Series No. 44A; WHO/Food Add/68.33).

16. FAO/WHO (1969) Specifications and criteria for identity and purity of some flavouring substances and non-nutritive sweetening agents (FAO Nutrition Meetings Report Series No. 44B; WHO/Food Add/69.31).

17. FAO/WHO (1969) Specifications for the identity and purity of food additives and their toxicological evaluation: some antibiotics. Twelfth report of the Joint FAO/WHO Expert Committee on Food Additives (FAO Nutrition Meetings Report Series No. 45; WHO Technical Report Series No. 430).

18. FAO/WHO (1969) Specifications for the identity and purity of some antibiotics (FAO Nutrition Meetings Report Series No. 45A; WHO/Food Add/69.34).

19. FAO/WHO (1970) Specifications for the identity and purity of food additives and their toxicological evaluation: some food colours, emulsifiers, stabilizers, anticaking agents, and certain other substances. Thirteenth report of the Joint FAO/WHO Expert Committee on Food Additives (FAO Nutrition Meetings Report Series No. 46; WHO Technical Report Series No. 445).

20. FAO/WHO (1970) Toxicological evaluation of some food colours, emulsifiers, stabilizers, anticaking agents, and certain other substances (FAO Nutrition Meetings Report Series No. 46A; WHO/Food Add/70.36).

21. FAO/WHO (1970) Specifications for the identity and purity of some food colours, emulsifiers, stabilizers, anticaking agents, and certain other food additives (FAO Nutrition Meetings Report Series No. 46B; WHO/Food Add/70.37).

22. FAO/WHO (1971) Evaluation of food additives: specifications for the identity and purity of food additives and their toxicological evaluation: some extraction solvents and certain other substances; and a review of the technological efficacy of some antimicrobial agents. Fourteenth report of the Joint FAO/WHO Expert Committee on Food Additives (FAO Nutrition Meetings Report Series No. 48; WHO Technical Report Series No. 462).

23. FAO/WHO (1971) Toxicological evaluation of some extraction solvents and certain other substances (FAO Nutrition Meetings Report Series No. 48A; WHO/Food Add/70.39).

24. FAO/WHO (1971) Specifications for the identity and purity of some extraction solvents and certain other substances (FAO Nutrition Meetings Report Series No. 48B; WHO/Food Add/70.40).

25. FAO/WHO (1971) A review of the technological efficacy of some microbial agents (FAO Nutrition Meetings Report Series No. 48C; WHO/Food Add/70.41).

26. FAO/WHO (1972) Evaluation of food additives: some enzymes, modified starches, and certain other substances: toxicological evaluations and specifications and a review of the technological efficacy of some antioxidants. Fifteenth report of the Joint FAO/WHO Expert Committee on Food Additives (FAO Nutrition Meetings Report Series No. 50; WHO Technical Report Series No. 488).

27. FAO/WHO (1972) Toxicological evaluation of some enzymes, modified starches, and certain other substances (FAO Nutrition Meetings Report Series No. 50A; WHO Food Additive Series No. 1).

28. FAO/WHO (1972) Specifications for the identity and purity of some enzymes and certain other substances (FAO Nutrition Meetings Report Series No. 50B; WHO Food Additive Series No. 2).

29. FAO/WHO (1972) A review of the technological efficacy of some antioxidants and synergists (FAO Nutrition Meetings Report Series No. 50C; WHO Food Additive Series No. 3).

30. FAO/WHO (1972) Evaluation of certain food additives and the contaminants mercury, lead, and cadmium. Sixteenth report of the Joint FAO/WHO Expert Committee on Food Additives (FAO Nutrition Meetings Report Series No. 51; WHO Technical Report Series No. 505 and corrigendum).

31. FAO/WHO (1972) Evaluation of mercury, lead, cadmium, and the food additives amaranth, diethylpyrocarbamate, and octyl gallate (FAO Nutrition Meetings Report Series No. 51A; WHO Food Additives Series No. 4).

32. FAO/WHO (1974) Toxicological evaluation of certain food additives with a review of general principles and of specifications. Seventeenth report of the Joint FAO/WHO Expert Committee on Food Additives (FAO Nutrition Meetings Report Series No. 53; WHO Technical Report Series No. 539 and corrigendum) (out of print).

33. FAO/WHO (1974) Toxicological evaluation of certain food additives including anticaking agents, antimicrobials, antioxidants, emulsifiers, and thickening agents (FAO Nutrition Meetings Report Series No. 53A; WHO Food Additives Series No. 5).

34. FAO/WHO (1978) Specifications for the identity and purity of thickening agents, anticaking agents, antimicrobials, antioxidants, and emulsifiers (FAO Food and Nutrition Paper No. 4).

35. FAO/WHO (1974) Evaluation of certain food additives. Eighteenth report of the Joint FAO/WHO Expert Committee on Food Additives (FAO Nutrition Meetings Report Series No. 54; WHO Technical Report Series No. 557 and corrigendum).

36. FAO/WHO (1975) Toxicological evaluation of some food colours, enzymes, flavour enhancers, thickening agents, and certain other food additives (FAO Nutrition Meetings Report Series No. 54A; WHO Food Additive Series No. 6).

37. FAO/WHO (1975) Specifications for the identity and purity of some food colours, flavour enhancers, thickening agents, and certain food additives (FAO Nutrition Meetings Report Series No. 54B; WHO Food Additives Series No. 7).

38. FAO/WHO (1975) Evaluation of certain food additives: some food colours, thickening agents, smoke condensates, and certain other substances. Nineteenth report of the Joint FAO/WHO Expert Committee on Food Additives (FAO Nutrition Meetings Report Series No. 55; WHO Technical Report Series No. 576).

39. FAO/WHO (1975) Toxicological evaluation of some food colours, thickening agents, and certain other substances (FAO Nutrition Meetings Report Series No. 55A; WHO Food Additive Series No. 8).

40. FAO/WHO (1976) Specifications for the identity and purity of certain food additives (FAO Nutrition Meetings Report Series No. 55B; WHO Food Additive Series No. 9).

41. FAO/WHO (1976) Evaluation of certain food additives. Twentieth report of the Joint FAO/WHO Expert Committee on Food Additives (FAO Food and Nutrition Series No. 1; WHO Technical Report Series No. 599).

42. FAO/WHO (1976) Toxicological evaluation of certain food additives (WHO Food Additives Series No. 10).

43. FAO/WHO (1977) Specifications for the identity and purity of some food additives (FAO Food and Nutrition Series No. 1B; WHO Food Additive Series No. 11).

44. FAO/WHO (1978) <u>Evaluation of certain food additives. Twenty-first report of the Joint FAO/WHO Expert Committee on Food Additives</u> (WHO Technical Report Series No. 617).

45. FAO/WHO (1977) <u>Summary of toxicological data of certain food additives</u> (WHO Food Additives Series No. 12).

46. FAO/WHO (1977) <u>Specifications for the identity and purity of some food additives, including antioxidants, food colours, thickeners, and others</u> (FAO Nutrition Meetings Report Series No. 57).

47. FAO/WHO (1978) <u>Evaluation of certain food additives and contaminants. Twenty-second report of the Joint FAO/WHO Expert Committee on Food Additives</u> (WHO Technical Report Series No. 631).

48. FAO/WHO (1978) <u>Summary of toxicological data of certain food additives and contaminants</u> (WHO Food Additive Series No. 13).

49. FAO/WHO (1978) <u>Specifications for the identity and purity of certain food additives</u> (FAO Food and Nutrition Paper No. 7).

50. FAO/WHO (1980) <u>Evaluation of certain food additives. Twenty-third report of the Joint FAO/WHO Expert Committee on Food Additives</u> (WHO Technical Report Series No. 648 and corrigenda).

51. FAO/WHO (1980) <u>Toxicological evaluation of certain food additives</u> (WHO Food Additive Series No. 14).

52. FAO/WHO (1979) <u>Specifications for the identity and purity of food colours, flavouring agents, and other food additives</u> (FAO Food and Nutrition Paper No. 12).

53. FAO/WHO (1980) <u>Evaluation of certain food additives. Twenty-fourth report of the Joint FAO/WHO Expert Committee on Food Additives</u> (WHO Technical Report Series No. 653).

54. FAO/WHO (1980) <u>Toxicological evaluation of certain food additives</u> (WHO Food Additives Series No. 15).

55. FAO/WHO (1980) <u>Specifications for the identity and purity of food additives (sweetening agents, emulsifying agents, and other food additives)</u> (FAO Food and Nutrition Paper No. 17).

56. FAO/WHO (1981) Evaluation of certain food additives. Twenty-fifth report of the Joint FAO/WHO Expert Committee on Food Additives (WHO Technical Report Series No. 669).

57. FAO/WHO (1981) Toxicological evaluation of certain food additives (WHO Food Additives Series No. 16).

58. FAO/WHO (1981) Specifications for the identity and purity of food additives (carrier solvents, emulsifiers and stabilizers, enzyme preparations, flavouring agents, food colours, sweetening agents, and other food additives (FAO Food and Nutrition Paper No. 19).

59. FAO/WHO (1982) Evaluation of certain food additives and contaminants. Twenty-sixth report of the Joint FAO/WHO Expert Committee on Food Additives (WHO Technical Report Series No. 683).

60. FAO/WHO (1982) Toxicological evaluation of certain food additives (WHO Food Additives Series No. 17).

61. FAO/WHO (1982) Specifications for the identity and purity of certain food additives (FAO Food and Nutrition Paper No. 25).

62. FAO/WHO (1983) Evaluation of certain food additives and contaminants. Twenty-seventh report of the Joint FAO/WHO Expert Committee on Food Additives (WHO Technical Report Series No. 696 and corrigenda).

63. FAO/WHO (1983) Toxicological evaluation of certain food additives and contaminants (WHO Food Additives Series No. 18).

64. FAO/WHO (1983) Specifications for the identity and purity of certain food additives (FAO Food and Nutrition Paper No. 28).

65. FAO/WHO (1983) Guide to specifications - General notices, general methods, identification tests, test solutions, and other reference materials (FAO Food and Nutrition Paper No. 5, Rev. 1).

66. FAO/WHO (1984) Evaluation of certain food additives and contaminants. Twenty-eighth report of the Joint FAO/WHO Expert Committee on Food Additives (WHO Technical Report Series No. 710).

67. FAO/WHO (1984) Toxicological evaluation of certain food additives and contaminants (WHO Food Additives Series No. 19).

68. FAO/WHO (1984) Specifications for the identity and purity of food colours (FAO Food and Nutrition Paper No. 31/1).

69. FAO/WHO (1984) Specifications for the identity and purity of food additives (FAO Food and Nutrition Paper No. 31/2).

70. FAO/WHO (1986) Evaluation of certain food additives and contaminants. Twenty-ninth report of the Joint FAO/WHO Expert Committee on Food Additives (WHO Technical Report Series No. 733).

71. FAO/WHO (1986) Toxicological evaluation of certain food additives and contaminants (WHO Food Additives Series No. 20).

72. FAO/WHO (1986) Specifications for the identity and purity of certain food additives (FAO Food and Nutrition Paper No. 34).

INDEX

Acetic anhydride 37
Acyl CoA oxidase 149-150
ADI
  calculation of 79, 96-97, 114
  definition of 75, 111
  group 27, 82-83, 91, 97
  not specified 83-84, 111, 136
  temporary 36, 84, 114, 147
Adrenal medullary lesions 44
Aflatoxins 28
Alkylation 148
Allergic reactions 21, 73-74, 92, 99, 112
Anaphalaxis 72-73, 112
Anorexia, toxic 82
Anticaking agents 37
Antifoaming/defoaming agents 37
Antinutritional factors 67, 94
Antioxidants 33, 37
Aromatic amines 28
Arsenic 87
Aspartame 33

Behavioural effects 40, 64
Biological versus statistical significance 45, 129
Bladder tumours 41, 43, 80, 146-147
Blood-brain barrier 60, 62
Body-weight gain, effects on 40, 78, 82, 113, 117
Bulking agents 21, 97

Caecal enlargement 41, 59, 67, 82
Cadmium 86-87
Calcium 41-42, 67
Caloric balance 82, 95
Capsaicin 89
Carbamates 28
Carcinogenesis
  secondary mechanism 53, 80-81
  transplacental 46, 64, 114
Carcinogenic
  contaminants 54, 64, 87-88
  food additives 42-46, 53, 80-81
Carcinogenicity, prediction of 28, 49
Carcinogens, classification of 45
Cardiomyopathy, beer drinkers 72

Cardiovascular lesions 39
Carrier solvents 43, 140
Catalase 150
Chinese Restaurant Syndrome 73
Clarifying agents 37
Cobalt 72
Codex Alimentarius Commission 17, 23-24, 38, 110-111
Codex Committee on Food Additives 29, 38, 110-111
Colours, food 41, 43, 137-138
Commission of the European Communities 63
Committee on Safety of Medicines 47, 110
Committee on Toxicity 47, 110
Contaminants 86-88
  carcinogenic 54, 64, 87-88
  estimating intake of 27, 84-85
  exposure of infants and young children to 61-65
  irreducible level 87, 112
  in processing aids 137-140
  provisional maximum tolerable daily intake 87-88, 110, 113
  provisional tolerable weekly intake 65, 87, 111, 114
  secretion in breast milk 64-65
  specifications for 32-35, 91, 93-94, 97
Control groups
  concurrent 44
  historical 44, 123
Council of Europe 89
p-Cresol 59
Cross fostering 82
Cyclamate 145-146
Cyclohexylamine 145

DEAE-cellulose 137
Decolourizing agents 37
Department of Health and Social Security 98
Detoxification of ingested compounds 41, 51, 62
Development, effects on 46-48, 61, 69, 80, 112
Diet, nutritional adequacy of 66-68, 93-95, 97-98
Diethylstilboestrol 64

Embryotoxicity 46-49, 80, 112
Enterohepatic circulation 51, 112
Environmental Protection Agency 47, 110
Enzyme
  deficiencies 73
  development 60-61, 64-65, 80
  immobilizing agents 34, 37, 86, 136-137
  induction 41, 53, 62, 82, 150

Enzyme (contd)
  preparations 18, 34, 37, 86, 135-137
Epidemiological studies 71-73, 122, 141
Epoxides 28
Estragole 151
Ethylenimine 34
Ethylnitrosourea 64
European Economic Community 21, 90, 92, 110
Exposure
  estimates 26-27, 30, 79, 84-85, 91
  patterns 22-23, 27, 76
Extraction solvents 18, 34, 36-37, 86, 138-140, 147-151

Fetotoxicity 46-48, 64-65, 112
Flavouring agents 16, 30, 56, 83, 86, 88-92
  criteria for testing 16, 88-92
  extracts 89, 91
  natural occurrence 88-91
Flora, intestinal 51, 53, 56-60, 80, 96, 145-146
  adaptation by 53, 57-58, 96
  age-related 60
  antibacterial activity of 59
  effects of antibiotics on 58
Food additive(s)
  in baby food 61
  background occurrence in food 84-85, 111
  carcinogenic 42-46, 53, 80-81
  chemical composition of 32-33, 93-94, 98
  criteria for testing 16, 25-29
  definition 18
  exposure of infants and young children to 60-63, 97
  in medical foods 69, 97
  natural occurrence 16, 25, 28-29, 75, 79, 87, 93
  normal body constituents, metabolism into 29, 54-56, 93, 96-97
  reactivity of 33-34, 51, 56, 66-68, 94, 139
  secretion in breast milk 63
Food and Drug Administration 47, 104, 110
Food processing 33-38, 56, 66, 94
Formaldehyde 148

Gastric papillomas 44
Gavage 53, 148-150
Germ-free animals 57
Global Environmental Monitoring System 27, 110
Glutamate 62, 73
Glutaraldehyde 37, 137

Glutathione 51, 148-149
Good laboratory practice 30-31, 110, 120
Good manufacturing practice 34-36, 84
Gums, plant 41

Haemagglutinins 94
Halogenated hydrocarbons 65, 139-140, 147-151
Health status of experimental animals 68
Hepatocellular carcinoma 149-150
Hepatocytes 150
Hepatoma 17, 44
Human data 21, 29, 40, 70-75, 89, 92, 97-99, 122, 141
Human use, history of 29
Hybridization 98
Hydrogen peroxide 76, 149-150
Hypersensitivity 71, 89, 99

Idiosyncratic reactions 71, 89
Immunoassays 74
Immunoglobulin E 73-74
Immunosuppression 39
Indoxyl sulfate 146
International Agency for Research on Cancer 21, 45, 64, 90, 110, 117
International Programme on Chemical Safety 47, 63, 110
Intolerance, food 29, 40, 73-75, 92, 99, 112
In utero studies 63-65
In vitro studies 48-50, 54, 57
Irreducible level 87, 112
Iron 81

Lactose 41
Laxative effects 40, 82, 85
Lead 86-87
Leukocyte histamine release assay 74
Liver
    enlargement 41, 82
    neoplasms 17, 44, 148-150
Lubricants/release agents 37
Lung neoplasms 44, 68, 148
Lymphomas 17, 44

Magnesium deficiency 42
Marketing trials 99
Maternal toxicity 47, 64

Maximum tolerated dose 40, 110, 113, 117, 148
Menthol 89
Mercury 64, 86-87
Metabolism and pharmacokinetics 28-29, 50-59, 70, 78, 80, 82, 91, 95-96, 138, 145-151
  absorption 29, 50-53, 57, 68, 70, 138, 141
  disposition 15, 29, 51-52, 70, 138
  elimination 50-51, 68, 70, 112, 138, 141
  interspecies variation 51-53, 60, 78
  into normal body consituents 29, 54-56, 93, 96-97
Metals in food 18, 33-34, 67-68, 86-88, 94, 97, 114, 141-142
Methylene chloride 147-149
Microencapsulation 137
Microsomal enzymes 41, 82, 148, 150
Migration studies 142-143
Mineral oil 42
Ministry of Agriculture, Forestry and Fisheries 47, 110
Modified starches 37, 41, 92
Mutagenicity 16, 28, 46, 49, 90, 92, 148
Mycelia, fungal 97
Mycotoxins 28, 86-87, 94, 97

Nephrocalcinosis 41-42, 59, 67
Nicotine 94
Nitrate 60
Nitrite 72
Nitrosamines 28, 64
No-observed-effect level, determination of 77-78, 117
Novel foods, 18, 21, 33, 92-99
Nutrients, essential 81, 87-88, 94-95, 141-142
Nutrition, mal- 59-60, 66
Nutritional deficiencies 42
Nutritional requirements, interference with 67, 94
Nutritive food ingredients 18, 69, 78, 81, 94-99, 117

Odour/taste-removing agents 37
Oil of sassafras 89, 91
Oil of wormwood 91
Oils, essential 89, 91
Oleoresins 89
Oncogenes 44
Organization for Economic Cooperation and Development 21, 47, 110

Packaging materials 17-18, 34, 43, 76, 86, 88, 142-143
Paired feeding studies 42, 82, 95

Palatability, lack of 78, 82, 95
Pancreatic adenomas 44
Parenteral administration 57
PCBs 64-65
Periodic review 19, 21-24, 36
Peroxisome proliferation 149-151
Pesticide residues 20, 141, 147
Pesticides Safety Precautions Scheme 47, 111
Phenol 57, 59
o-Phenylphenol 147
Pheochromocytomas 17, 44
Phosphorus 42, 67
Physiological responses 40-42, 78, 82-83
Phytate 94
Placental barrier 63-65, 68-69
Polycyclic aromatic hydrocarbons 28
Polyethylenimine 34, 137
Polyols 41, 92
Predicting toxicity 27-28, 30, 48-50, 91
Priority setting 25, 27, 29-30, 90-92
Processing aids
  anticaking agents 37
  antifoaming agents 37
  carrier solvents 43, 140
  clarifying agents 37
  decolourizing agents 37
  definition 113
  enzyme preparations 18, 34, 37, 86, 135-137
  extraction solvents 18, 34, 36-37, 86, 138-140, 147-151
  lubricants/release agents 37
  odour/taste-removing agents 37
Protein Advisory Group 98
Provisional Maximum Tolerable Daily Intake 87-88, 110, 113
Provisional Tolerable Weekly Intake 65, 87, 111, 114

Range-finding studies 77
Receptor sites 54, 89
Recombinant DNA 98
Renal weight changes 82
Reproduction
  studies 42, 47, 60-63, 80, 82
  toxicity 39, 46-47, 49, 69, 114
Risk management 76

Saccharin 41, 72-73, 146-147
Safety factor 75-76, 78-81, 84, 96-97, 114
Saffrole 89, 91

Sarcoma
  salivary gland 148
  subcutaneous 16
Scientific Committee on Problems of the Environment 47, 111
Senescence 65
Skin test 74
Sorbitol 92
Specifications 32-38, 93-94
  Codex advisory 38
  for complex mixtures 32, 93
Stabilizers 37, 139-140, 149
Statistical analysis 115-134
  age adjustment 127
  analysis of covariance 132
  animal placement 123
  Armitage test 131
  association between variables 128, 132
  Bartlett test 132
  between-animal comparisons 131-132
  between-group comparisons 126-128
  bias 115-116, 121-124
  blocking factor 121
  chance 115-116, 122, 129-130
  Cochran test 131
  contingency table analysis 125
  continuous data 125-126, 130, 132
  crossed design 120
  dose-related trend 126
  dose response 45, 48, 77-78, 81, 117
  duration of study 65, 69, 117, 120
  experimental unit 124-125
  Fisher exact test 131
  Friedman two-way analysis of variance 131
  heterogeneity 126, 131-132
  hypothesis testing 129
  Kruskal-Wallis one-way analysis of variance 131
  linear regression analysis 132
  Mann Whitney U test 131
  McNemar test 131
  multiple comparisons 130
  multiple observations 125, 128
  non-parametric analysis 130
  number of animals 23, 118, 120, 127-128
  observational units 124-125
  one-way analysis of variance 131-132
  over-responsive species 17, 43-45, 69, 116
  $\underline{P}$-value 129-130
  page test 131
  paired t-test 132

Statistical analysis (contd)
　parametric methods 125
　Pearson correlation coefficient 132
　precision 118, 120, 122
　presence/absence data 125, 130-131
　probability values 129
　randomization 122-124
　ranked data 125
　response variable 125, 127
　Spearman's rank correlation coefficients 132
　stratification 121, 123, 126-127
　stratified random sampling 122
　stratified tests 131
　students t-test 132
　test for heterogeneity 126
　test for outliers 132
　two-way analysis of variance 131-132
　types I and II error 118
　Wilcoxon matched pairs signed-rank test 131
　within-animal comparisons 128, 131-132
Structure/activity relationships 27-29, 83, 90-92
Styrene 87
Sucrose esters of fatty acids 56
Sulfites 67, 73
Sweeteners, bulk 83

Tansy 91
Tartrazine 73
Teratogenicity 46-48, 80
Testicular atrophy 42, 145-146
Tetrachloroethylene 139
Thiamine 67
Thujone 91
Tin 87
Trichloroacetic acid 150-151
Trichloroethane 139
Trichloroethylene 139, 149-151
Trypsin inhibitors 94
Tryptophan metabolites 146
Tumours
　benign 44-45
　bladder 41, 43, 80, 146-147
　commonly-occurring 17, 43-45, 69, 116
　endocrine-associated 44
　liver 17, 44, 148-150
　lung 44, 68, 148
　malignant 44-45
　mammary 44

Tumours (contd)
  in old age 65, 69, 120
  pancreatic 44
  pituitary 44
  thyroid 44

Veterinary drug residues 17-18, 21, 42, 86, 140-141
  anabolic agents 17, 42, 86, 140-141
  growth promoters 18

Xylitol 92

Zingiberin 89